Reliability in Instrumentation and Control

Industrial Instrumentation Series
Series Editor: Dr B E Noltingk
Published with the Institute of Measurement and Control

The Industrial Instrumentation Series of books describes, authoritatively and in detail, the instrumentation available in the major industries, and the ways in which instruments are put to use. Classical instruments and the latest applications of new technology in the field are covered. Emphasis is placed on those aspects of an industry (such as measurements required or environmental constraints) that particularly influence the instrumentation used.

Each volume is written by experienced practitioners with specialist knowledge of instruments for particular applications. Each book contains extensive lists of references. Thus every title provides a comprehensive and profound treatment never before offered on the practical use of instrumentation. This makes the series essential for anyone involved in the specification, design, supply, installation, use, maintenance and repair of instruments.

The series will cover all major industries in which instrumentation is used and all important issues relating to the use of instruments in industry.

The Institute of Measurement and Control was founded in the UK in 1944 as the Society of Instrument Technology and took its present name in 1968. It was incorporated by Royal Charter in 1975 with the object 'to promote for the public benefit by all available means the general advancement of the science and practice of measurement and control technology and its application'. The Institute of Measurement and Control provides routes to Engineering Council status as Chartered and Incorporated Engineers and Engineering Technicians. The Institute is the UK member organization of the International Measurement Confederation (IMEKO).

Titles in the Industrial Instrumentation Series

Analytical Instrumentation for the Water Industry
T R Crompton
0 7506 1139 1   324pp   1991

Instrumentation and Sensors for the Food Industry
Edited by E Kress-Rogers
0 7506 1153 7   450pp   1992

Power Station Instrumentation
Edited by M W Jervis
0 7506 1196 0   700pp   1993

Flow, Level and Pressure Measurement in the Water Industry
G Fowles
0 7506 1047 6   240pp   1993

Reliability in Instrumentation and Control

J. C. CLULEY, MSc, CEng, MIEE, FBCS

Series Editor: B. E. NOLTINGK

Published in association with The Institute of Measurement and Control

Butterworth-Heinemann Ltd
Linacre House, Jordan Hill, Oxford OX2 8DP

 PART OF REED INTERNATIONAL BOOKS

OXFORD LONDON BOSTON
MUNICH NEW DELHI SINGAPORE SYDNEY
TOKYO TORONTO WELLINGTON

First published 1993

© Butterworth-Heinemann Ltd 1993

All rights reserved. No part of this publication
may be reproduced in any material form (including
photocopying or storing in any medium by electronic
means and whether or not transiently or incidentally
to some other use of this publication) without the
written permission of the copyright holder except in
accordance with the provisions of the Copyright,
Designs and Patents Act 1988 or under the terms of a
licence issued by the Copyright Licensing Agency Ltd,
90 Tottenham Court Road, London, England W1P 9HE.
Applications for the copyright holder's written permission
to reproduce any part of this publication should be addressed
to the publishers

British Library Cataloguing in Publication Data
Cluley, J. C.
  Reliability in Instrumentation and
  Control. – (Industrial Instrumentation
  Series)
  I. Title II. Series
  629.8

ISBN 0 7506 0737 8

Library of Congress Cataloguing in Publication Data
Cluley, J. C. (John Charles)
  Reliability in instrumentation and control/J. C. Cluley.
    p. cm. – (Industrial instrumentation series)
  Includes bibliographical references and index.
  ISBN 0 7506 0737 8
  1.Engineering instruments – Reliability. 2. Automatic control –
Reliability. I. Title. II. Series.
TA165.C577   92–21340
629.8–dc20   CIP

Composition by Genesis Typesetting, Rochester, Kent
Printed in Great Britain by Redwood Press Limited, Melksham, Wiltshire

Contents

Preface — vii

1 Introduction to control systems — 1

2 Reliability principles and terminology — 11

3 Reliability assessment — 23

4 System design — 41

5 Building high-reliability systems — 55

6 The human operator in control and instrumentation — 67

7 Safety monitoring — 79

8 Software reliability — 87

9 Data transmission — 105

10 Electronic and avionic systems — 113

11 Nuclear reactor control systems — 125

12 Process and plant control — 139

Index — 153

Preface

The increasing use of automatic control systems in situations where failure can involve a serious risk to human life brings with it a demand for very high reliability. This requirement extends throughout the system from the transducers and measuring devices which establish the current state of the system to the output devices which regulate the system. Consequently there is a growing need for engineers involved in the specification, design, construction and maintenance of these systems to understand the need for reliability and the factors which affect it.

There is also a growing public awareness of the risks which accompany certain activities such as the generation of electricity using nuclear reactors and the processing of dangerous chemicals. This has been reinforced by the publicity following the Flixborough accident in the UK, the Three Mile Island nuclear accident in Harrisburg, USA, and the near destruction of the nuclear reactor at Chernobyl in the Ukraine.

In all of these accidents human operators were part of the control activity and their errors and confusion had no small effect on the final result.

In this text I have tried to encompass all of the main factors which affect reliability in instrumentation and control systems and measures such as the introduction of redundancy which can enhance reliability. Since so many systems now include embedded computers it is essential to consider also the reliability of the software which controls them.

I hope that the book will be found useful by all engineers involved in any way with these systems. As all degree courses in engineering should now contain some treatment of system and component reliability, I hope that the book will also be useful for students on all such courses.

1 Introduction to control systems

1.1 Instrumentation and control systems

The *Encyclopaedia Britannica* defines an instrument as 'that which can be used as a means to an end, hence a mechanical contrivance, implement or tool'. This definition is too wide for the purposes of this book, since it includes, for example, both musical and surgical instruments. We are concerned only with instruments used in science or engineering, nearly all of which are required to measure some state or quantity. This may be recorded to provide evidence of the output of a plant or process to satisfy the plant operator or the customer, or, as with tachygraph discs, to satisfy legal requirements. Such instruments are also widely used in control systems since an essential precursor to an attempt to control any system or variable is some means of measuring the state of the system or the value of the variable. Most control systems are closed-loop or feedback systems in which the current state of the system is measured and compared with the desired state. The difference between the two states is used as an input to

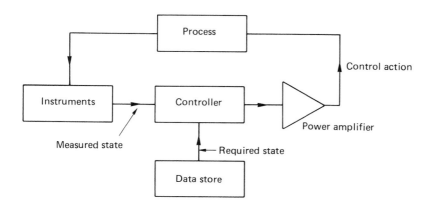

Figure 1.1 Block diagram of typical control system

modify the current state to bring it nearer to that desired. The general arrangement is shown in Figure 1.1. A typical system may include data storage and transmission, signal processing and power amplification, but an essential component is some form of instrumentation which can measure the current state of the system.

Although many early instruments such as thermometers and pressure gauges had their output scales built into the body of the instrument, later devices used in systems of increasing complexity were required to present their information at a point remote from the body of the instrument and from the quantity being measured. Thus plant and process control was concentrated in rooms specially set aside for this purpose, with all measured quantities being displayed to assist the operator. In large plant such as coal-fired power stations the data may have to be transmitted some hundreds of metres. Also the concentration of the control activity in one place means that many plant operations must be remotely controlled. There is consequently a need for both data and control signals to be transmitted to a remote site.

Such transmission can be performed by mechanical means (such as the steel cables formerly used to operate railway signals) or by pneumatic systems (as used in some boiler houses and chemical plants), but the majority of installations now use electrical signals for transmitting both measurement and control information. This has the advantage of cheapness and convenience, particularly where data must be processed, for example by removing a zero offset or linearizing the response of a transducer. It also much simplifies the task of storing data which is needed in many processes to enable the operators to have a record of plant performance. They may thus determine whether the plant parameters have been kept within acceptable limits and also detect any significant changes in a particular variable.

1.2 The use of microprocessors

The trend in current instrumentation is consequently to convert all measured quantities into an equivalent electrical signal before processing, storing, displaying or printing them. This change has been accelerated by the use in instrumentation of microprocessors, which can directly handle only electrical input and output signals. Microprocessors are used not only for controlling a set of instruments in such systems as automatic test equipment, but also in individual instruments such as analytical devices where they may regulate automatic calibration runs. They are also used in some complex instruments such as high-performance cathode-ray oscilloscopes and logic state analysers where they may be programmed to perform a system test when power is first applied before allowing the user access to the system.

Although some early digital control systems used hardwired logic, this was very difficult to alter as requirements changed as individual customers required particular system features. The use of computers (in most cases in the form of microprocessors) means that the particular details of the task to be performed reside not in a mass of wiring between logic packages, but in the computer program. In order to allow operation as soon as the system is energized, the program is normally held in a 'read-only' store whose contents cannot be corrupted or changed in any way by the action of the computer.

If the computer task needs to be changed, this can be done very quickly by removing the old store package and plugging in a new one. If fairly regular changes are envisaged the store package can be an erasable one; a few minutes' exposure to UV light will then remove the stored data and a new program can be loaded into it.

A further advantage of using computers in control systems lies in the very fast computations which they can perform, so enabling complex control algorithms to be used and also operations such as pattern recognition which involve handling large volumes of data. The power of current microprocessors is such that they can cope with the volumes of data and the need for fast processing in many control systems, and if not the task can easily be subdivided and given to several microprocessors.

1.3 Data representation

In any but the simplest control or instrumentation system signals representing measured quantities are needed to supply display, recording, signal conversion or power output devices. For ease in assembling such systems it would be convenient for all signals to be compatible and have the same zero and full-scale values. Many transducers produce outputs which are analogue signals so that their magnitudes are related directly to the magnitudes of the physical quantities they represent. Analogue output signals are also needed to drive chart recorders and some display devices. Many devices use a range of 0 to 10 V for analogue signals; for maximum dynamic range this should be as large as possible but a choice of 10 V allows signal processing to be performed by cheap and compact integrated circuit amplifiers, most of which have an output voltage capacity of just over ± 10 V.

The limiting factor here is the bandwidth of the amplifiers and transmission lines which cannot deal adequately with frequencies above some tens of kilohertz. Any system which has to handle frequencies in the megahertz region and above will need the transmission lines to be correctly terminated, typically with an impedance of 50-200 ohms. Thus unless signals are limited to an amplitude of 1 V or less a great deal of power is required to drive lines and devices. Consequently high-speed devices which sample and convert, say, 10 million samples per second will generally have a full-scale voltage of only 1 V for analogue signals.

Using a voltage scaling for analogue signals is convenient for systems with short transmission runs and where electronic circuits can be used to avoid drawing appreciable currents from them. They are liable to introduce errors, however, where indicating instruments have to be driven from remote measurements some hundreds of metres away. These require appreciable current and there is a significant voltage drop in the transmission lines.

The method adopted to overcome this difficulty is to use a current source so that the resistance of the lines does not affect the current. This arrangement is generally used with a range of either 0-20 mA or 4-20 mA. The advantage of the latter is that an open-circuit line is immediately detected since it gives a current outside the permitted range.

A similar off-zero scaling is used in pneumatic control and instrumentation systems which use air pressure as the analogue variable which represents data values. For signal transmission zero is represented by a pressure of 20 kPa and the maximum value by 100 kPa. The corresponding pressures in Imperial units are 3 and 15 psi. This arrangement was developed for the petroleum and chemical industries where flammable vapours were often present and installations had to be intrinsically safe. Also there is no chance of picking up noise and interference from adjacent power cables and equipment, always a possible hazard when low-level electrical signals are transmitted. Many process control systems require considerable force to be exerted to open and close valves and operate other moving parts, and this is more easily provided by pneumatic rams and actuators than electric motors. Thus even when the rest of the system uses electrical signals the final output devices may be pneumatically operated.

The disadvantages include a time lag of around a second for each 100 m of pipeline and the inability to perform much signal conditioning. Also there is no possibility of data storage.

Although many transducers produce analogue outputs, any but the smallest control or instrumentation system will have to cope also with digitally coded signals since any computer embedded in the system can accept and output only these signals; they are also generated as inputs from keyboards and are needed for output devices such as CRT displays and numerical displays. Most control and instrumentation systems will generally have a mixture of analogue and digital components and will thus require analogue-to-digital converters (ADCs) and digital-to-analogue converters (DACs).

1.4 Data highway systems

The decreasing cost of modern integrated circuit electronic devices has meant that the cost of testing any but the simplest systems becomes a disproportionate part of the total device cost. The usual way to reduce the cost of testing is to construct automatic test rigs which can be programmed to perform the range of tests previously carried out manually. Where equipment is normally made in batches a high degree of flexibility in the test rig is needed, so that it can readily be altered to test a new product by assembling a different set of test instruments with a different control programme. This process is greatly simplified if all instruments and the controlling computer are fitted with identical compatible interfaces so that any required combination can quickly be connected together. Several attempts have been made to specify such a system.

The first arose from the activities of the committee for European Standards on Nuclear Equipment (ESONE). This specified a standard data highway system for all instruments and controllers used in nuclear research and the operation of nuclear power stations. It was called CAMAC and was first published in 1969. It was revised and extended in 1975 and this is the current version. It allowed for parallel data transfers of up to 24 bits with up to 16 address bits and 14 control bits. All units are fitted with an 86 way printed circuit edge connector.

The advantages of such a standard system were soon apparent outside Europe and the same arrangement was adopted in the USA and in the USSR. In many cases 16 bit computers were used so that eight of the data lines were not used. The amplitudes and timing of the signals were closely specified to ensure that units from different manufacturers and different countries would always work together without errors. In all many hundreds of instruments are available which are fitted with a standard CAMAC interface.

The high-speed and complex addressing and control features of the CAMAC system result in a complicated and rather expensive arrangement, and the need was perceived for a much simpler system which lent itself to the use of early 8 bit microprocessors for assembling less complex test rigs. The Hewlett-Packard Company of America proposed a scheme with eight data and eight control lines which could address 15 devices, called the General Purpose Interface Bus (GPIB). A later amendment allowed an extra control byte to be transmitted to control up to 31 different functions at each device. The data highway was bidirectional unlike the CAMAC scheme which uses two 24 bit unidirectional highways, one for input and the other for output.

The GPIB was soon used widely in North America and was adopted by the IEEE as the IEEE-488 bus, and later internationally as the IEC 625 bus. The IEEE and IEC buses are electrically identical, but the IEEE bus uses a 24 way stackable connector whereas the IEC bus uses a standard subminiature 25 way connector specified for modems and other data handling equipment.

1.5 Bridge and comparison methods

Most measurements can easily be arranged to deliver analogue or digital outputs automatically, but some measuring processes have traditionally involved manual operations which are not easy to automate. Typical of these are bridge methods such as the measurement of electrical resistance. By using the out-of-balance signal from a Wheatstone bridge to control a motor driving the movable bridge contact, it can be turned into a self-balancing bridge as shown in Figure 1.2. However, the search for the

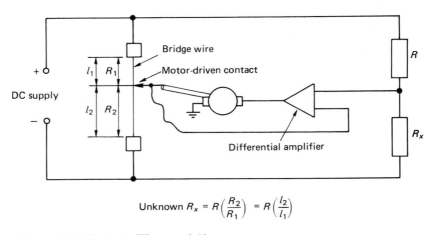

Unknown $R_x = R\left(\frac{R_2}{R_1}\right) = R\left(\frac{l_2}{l_1}\right)$

Figure 1.2 *Self-balancing Wheatstone bridge*

balance condition must be comparatively slow since it depends on mechanical movement. Where a rapid measurement is required the bridge technique is abandoned and a precision current generator is used to pass a known current through the component on test. The voltage across it can be measured by an ADC to give a digital signal suitable for direct input to a microprocessor.

The most accurate measurement of voltage is made by a potentiometer using comparison methods but this is generally too slow for many control systems which may need to sample a signal many thousands of times per second. This requires a high-speed sampling unit followed by a fast ADC.

Introduction to control systems 7

1.6 Display technologies

Most measurement systems are designed to indicate to an operator or user the result of their activity. The amount of information will vary widely, from the three decimal digit display for a personal weighing machine to the many graphs of sound amplitude against time displayed on a cathode-ray tube in three-dimensional format when testing the reverberation time of a room using a swept frequency pulsed tone. For simple numerical displays seven-segment light-emitting diodes (LEDs) are widely used. They can be obtained in sizes up to 25 mm high and have very long life expectation. They need only around 2 V to illuminate them, and so are conveniently driven by integrated circuit devices. Where a few characters of alphabetic information must be displayed dot matrix LEDs must be used.

Also available in similar sizes are liquid crystal displays (LCDs) which show black numerals against a light-grey background. These do not stand out visually as well as LEDs, but have the advantage of long life and require very little power, and so are more convenient for portable equipment running from batteries. They are readily available with provision for displaying up to four lines of 40 characters each having a 5×7 dot matrix format. Special LCD devices having considerably greater capacity are currently used for the screens of portable computers.

Another device with a bright display is the plasma panel which uses the glow discharge in a low-pressure gas. Initially a potential of about 180 V is needed to start conduction, falling to some 100 V to sustain it. The current causes an orange-coloured glow between the exciting electrodes which are usually two sets of perpendicular wires. These display panels can show up to 3000 characters. They require rather more complex driving circuits than LED displays on account of the much higher voltages needed, but these can easily be handled by the high-voltage transistors developed for television sets which are rated at 300-500 V.

Where it is required to display a large number of characters with upper and lower cases and a full selection of punctuation marks, including italic, subscript, superscript, etc., a CRT display is needed. This can typically display 25 lines, each having 80 characters. It produces a television-type raster display and needs a special character generation chip which converts the incoming ASCII-coded character into a series of pulses which when fed to the grid of the CRT will produce the required character on a 7×5 dot matrix. A high-resolution monochrome tube can display typically 640×400 points on the screen. This means that some 32 kbytes of storage are needed to hold the display information which must be read about 70 times per second to refresh the display. Colour displays usually have lower resolution, perhaps 320×200 points. If each point can have any one of 16 colours, 4 bits are needed for the colour information and the same storage volume of 32 kbytes will be needed to hold one screen of data. Colour displays are in demand for large control and monitoring systems such as those used for remotely controlled gas or oil pipelines. The layout of each pumping station along the route is stored in the controlling computer and can be called up in the event of some abnormal condition. In addition to the graphical information, values of flow, pressure and temperature, etc., can be shown on the screen.

Where measuring transducers produce analogue outputs the simplest form of display is some analogue device which can accept the transducer output directly, or with a simple electronic buffer to reduce transducer loading. This is generally a moving-pointer device; other methods involve a moving-scale or a bar-chart display. The latter can be a moving-ribbon device or a series of LEDs or LCDs which are progressively energized to generate a bar-type display. The length or height of the display is proportional to the magnitude of the variable concerned. Thirty-element LED bar displays are now readily available which permit a resolution of just over 3%. Large pointer-type instruments with mirror scales to reduce parallax errors can have accuracies of better than 1%, but those small enough to fit into the display panel of an aircraft or any similarly compact control panel will not have such high resolution. Thus any data display requiring a resolution of much better than 1% will have to use some form of digital output, or some special analogue technique such as the moving-scale display. This can have high resolution since there is no limit in principle to the length of the scale other than the size of the spools on which the scale is stored. However, the longer the scale the longer it takes to move from one end to the other and the slower the response.

Despite their limited resolution pointer instruments have advantages when it is necessary to monitor a number of instruments which should indicate similar values, for example the oil pressures, oil temperatures or vibration levels of the four engines in a passenger aircraft. A quick glance at the instruments will reveal that all four pointers showing, for example, the oil pressures are pointing in the same direction; this shows similar values for the four engines and is taken as a satisfactory condition. In the same way any indication of a disparity in the pressures shown by one pointer not being roughly parallel to the other three is very evident and should lead to a more detailed investigation. In order to obtain maximum contrast a black pointer against a white scale, or a white pointer against a black scale, is generally used. Some electronic analogue displays use LEDs arranged in a circular pattern, the value indicated being shown by the position of the one energized LED. This does not represent the appearance of a moving pointer very well and such displays are more likely to be misread.

1.7 Recording technologies

Many control and instrumentation systems require not only an indication of current conditions but also some storage of previous data to enable the long-term trends and variations to be studied. Where the measured quantities are changing comparatively slowly as in many chemical processes a chart recorder is a convenient way of retaining information; several traces can be recorded on the same chart so allowing the correlation between various quantities to be examined. The chart can be circular or, for long duration recording, in the form of a long roll. Most recorders can be driven at various chart speeds so that the degree of time resolution can be chosen to suit the rates at which the variables change.

The upper frequency limit of chart recorders is a few hertz; signals which vary more rapidly than this can be handled by a mirror galvanometer device which reflects a spot of light on to a moving strip of light-sensitive paper. By using ultraviolet (UV) light no chemical development is necessary, the trace becoming visible by exposing it to daylight. A spray is available to treat the chart if it is required as a permanent record. Up to 12 channels of data can be recorded on one chart. The highest frequency which the UV recorder can handle is a few kilohertz; frequencies above this can be recorded by using a cathode-ray tube as a light source.

One disadvantage of the pen-type chart recorder is that it requires a fixed mounting and cannot tolerate motion or vibration. Thus it is not acceptable for instrumentation systems which are used to collect data from motor vehicles or aircraft in flight. Magnetic tape recorders are often used for these applications since they can withstand the motions encountered. Direct recording of the signals is not usually acceptable since this does not retain the DC component, and the signals are usually encoded by frequency modulation so that the signal amplitude is indicated by the frequency of a sine wave recorded on the tape. The electronics are designed to ignore any amplitude variations which are inevitable due to slight variations in tape characteristics along its length and in tape-head contact. Fourteen channel recorders built to the IRIG standard are widely used in telemetry and aircraft testing. Magnetic recording is also used for the flight data recorders used on civil aircraft which store important flight parameters and are robustly built to survive with their information in the event of a crash. This application takes advantage of a valuable feature of magnetic recording – the ease with which data can be erased. In flight data recorders a continuous loop of the magnetic material can be used, the data being erased immediately before recording. Thus when a failure occurs and power disappears the data remaining will provide information about the state of the system leading up to the fault.

Magnetic and chart recorders are convenient for situations where the raw signal can be stored, but where some signal processing such as scale changing or a non-linear conversion is needed the fast arithmetic power of a computer and its associated semiconductor store must be invoked. In nearly all cases a microprocessor with a read-only program store meets the need and avoids reliance on a human operator. Where data is generated at high speed, as for example by a large wind tunnel, semiconductor storage is the only method of handling the high data rates.

Where a permanent record of numerical values, perhaps accompanied by text, is needed some form of printer is convenient. The cheapest and simplest device is a dot matrix printer which uses a vertical row of nine fine wires to print a dot on a roll of paper through an inked ribbon. The head holding the wires is moved across the paper by a small stepping motor and the impulses driving the wires are generated from a read-only store in the printer which converts from the ASCII-coded input signals to the pulses determined by the shape of the character being printed.

Typical speeds for dot matrix printers are 200 characters per second with 80 characters per line. Nine-pin printers show the dot structure; if this is obtrusive 24-pin printers are available whose output more closely resembles that of a typewriter. For the rare occasions when a much higher printing rate is needed line printers can be used. These print a complete line of output (up to 132 characters) in one operation and can run at speeds up to 2000 lines per minute.

Bibliography

Agard, P. J. *et al.* (1985) Information and display systems. In *Process Instruments and Control Handbook*, McGraw-Hill, New York

Barney, C. G. (1981) *Intelligent Instrumentation: Microprocessor Applications in Measurement and Control*, Prentice Hall, London

Bell, D. A. (1983) *Electronic Instrumentation and Measurements*, Reston Publishing, Reston, VA

Bentley, J. (1983) *Principles of Measurement Systems*, Longman, London

Carr, J. J. (1983) *Designing Microprocessor-based Instrumentation*, Reston Publishing, Reston, VA

Doebelin, E. O. (1983) *Measurement Systems Application and Design*, 3rd edn, McGraw-Hill, London

Gregory, B. A. (1981) *An Introduction to Electrical Instrumentation and Measurement Systems*, 2nd edn, Macmillan, London

Jones, E. B. *Jones' Instrument Technology*: Vol. 1 *Mechanical Measurements* (4th edn, 1985), Vol. 2. *Measurement of Temperature and Chemical Composition* (4th edn, 1985), Vol. 3 *Electrical and Radiation Measurements* (4th edn, 1987), Vol. 4 *Instrumentation Systems* (4th edn, 1987), Vol. 5 *Automatic Instruments and Measuring Systems* (1986), Butterworth, London

Morgan, C. G. (1983) *The Microcomputer in the Laboratory*, Sigma Technical Press, Wilmslow, Cheshire

Morrison, R. (1986) *Grounding and Shielding Techniques in Instrumentation*, Wiley, Chichester

Oliver, B. M. (1971) *Electronic Measurements and Instrumentation*, McGraw-Hill, London

Perez, R. A. (1988) *Electronic Display Devices*, TAB Books, Blue Ridge Summit, PA

Wilmshurst, T. H. (1985) *Signal Recovery from Noise in Electronic Instrumentation*, Hilger, Bristol

Wobschall, D. (1987) *Circuit Design for Electronic Instrumentation: Analog and Digital Devices from Sensor to Display*, 2nd edn, McGraw-Hill, London

2 Reliability principles and terminology

2.1 Definition of reliability

Reliability is generally defined as the probability that a component or assembly will operate without failure for a prescribed period under specified conditions. In order to ensure a meaningful result the conditions of operation, both physical and electrical, must be specified in detail, as must the standard of performance required.

Typical conditions which need to be considered are:

1 Variations in power supply voltage and size of voltage transients.
2 With AC supplies, the variations in frequency and harmonic content.
3 The level of unwanted RF energy radiated by the equipment must not cause interference to radio communications.
4 The equipment must be able to tolerate some RF radiation if it is to be used near high-power radio or radar transmitters.
5 Equipment for satellites and nuclear plants may need shielding from the ionizing radiations which they may experience.
6 Maximum and minimum ambient temperatures.
7 Maximum and minimum humidity.
8 Vibration and shock levels.
9 External conditions such as exposure to sand and dust storms, rainstorms, salt-water spray or solar radiation.
10 Air pressure.
11 Variations in loading (where relevant).

When quoting component reliability we need to consider only some of the above factors; for example, with small ceramic dielectric capacitors used in laboratory conditions only working voltage and temperature may be specified. When quoting the reliability of an electronic controller for the jet engine of an airliner, however, we may need to specify nearly all of the factors, since the operating conditions are much more severe.

Where a range of variation is shown, for example with temperature, the equipment may be subject to rapid changes or regular cycling and facilities must be provided to implement these changes when performing life tests. Also, to ensure testing under the worst conditions likely to be encountered it may be necessary to correlate variations of two parameters. For example, the worst conditions for producing condensation are minimum temperature and maximum humidity.

For continuously operating equipment the factor which indicates the amount of use is simply the elapsed time, but for intermittently used systems some deterioration may occur even though it is not energized. Thus the usage is indicated by adding to the switched-on time a fraction of the standby time. For some devices such as switches and relays usage is related to the number of operations rather than the switched-on time.

2.2 Reliability and MTBF

Although reliability as defined above is an important parameter to equipment users it has the disadvantage that its value depends upon the operating period so that the manufacturer cannot specify a value for reliability which applies to all applications. It is convenient to look for such a value; the one usually adopted is the mean time between failures (MTBF). This applies to a maintained system in which each failure is repaired and the equipment is then restored to service. For systems which cannot be maintained, such as satellite control systems, the value of interest is the mean time to failure (MTTF).

For maintained systems the MTBF can be measured by operating them for a period T hours long enough to produce a number N of faults. The MTBF is then $M = T/N$ hours. This expression assumes that the time between faults does not change significantly with time, a reasonable assumption for much electronic equipment. It is not possible to test every system for all of its life, so any value for MTBF is subject to a sampling error.

Another measure of reliability which is often quoted is failure rate, that is the number of failures per unit time. This is particularly useful in relation to individual components since it is the most convenient figure to use when estimating the reliability of a complete system in terms of the performance of its components. It is equal to the reciprocal of the MTBF, that is the failure rate $\lambda = 1/M$.

For electronic components which have inherently high reliability the time interval is usually taken as 10^6 or 10^9 hours; for smaller periods the failure rate will be a small fraction.

Equipment manufacturers generally quote MTBF as an indication of reliability since the value does not depend upon the operating period which is often different for each user. However, reliability is the factor of prime concern to the user and the relation between this and MTBF is thus important. The simplest case occurs when the failure rate and thus MTBF is constant with time. This is a somewhat drastic assumption but failure records of many maintained electronic systems show it to be justified. It is in most cases a slightly pessimistic assumption since the components used to replace faulty ones have been manufactured later after more experience with the production process has been acquired and are thus usually more reliable. Also those components which fail early are less reliable and in time are replaced by better versions.

Mechanical components can initially be assigned a constant failure rate but they will ultimately show an increasing failure rate as they reach the wear-out phase. Thus the maintenance programme will usually involve changing or refurbishing these items well before they reach wear-out.

The relation between MTBF and reliability can be established using the Poisson probability distribution. This applies to a random process where the probability of an event does not change with time and the occurrence of one event does not affect the probability of another event. The distribution depends on only one parameter, the expected number of events μ in the period of observation T.

The series

$$S = \exp(-\mu)\,(1 + \mu + \mu^2/2! + \mu^3/3! + \ldots + \mu^r/r! \ldots)$$

gives the probability that 0, 1, 2, 3 ... ,r events will occur in the period T. If we take an event to be a system failure the probability we require is that of zero failures, that is the reliability. This is then

$$R = \exp(-\mu) \tag{2.1}$$

2.3 The exponential failure law

If the system MTBF is M we expect one fault in the period M and thus T/M in the period T. T/M is thus the expected number of events μ and the reliability is

$$R = \exp(-T/M) \tag{2.2}$$

The two quantities T and M must be expressed in the same units. Graphs of this relation on linear and logarithmic scales are shown in Figures 2.1 and 2.2.

As an example we consider the instrumentation module of a research satellite which is designed to have a working life of 5 years. If it has an MTBF of 40 000 hours what is its probability of surviving 5 years without failure? Here T must be expressed in hours as $5 \times 8760 = 43\,800$.

14 *Reliability in Instrumentation and Control*

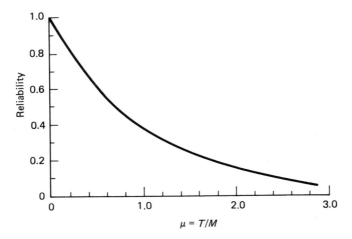

Figure 2.1 *Reliability–time graph – linear scale*

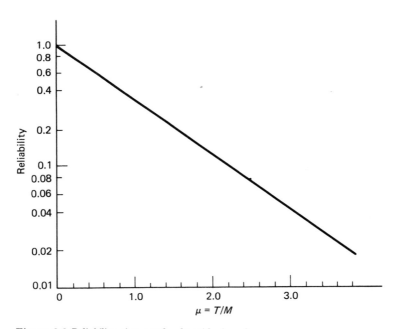

Figure 2.2 *Reliability–time graph – logarithmic scale*

Thus the probability of zero failures is

$\exp(-T/M) = \exp(-43\,800/40\,000) = 0.335$

This is not an acceptable figure. We can also use equation (2.2) to calculate what MTBF would be required to afford a more acceptable reliability, say 0.8. This gives

$0.8 = \exp(-43\,800/M)$

Thus

$-43\,800/M = \log_e (0.8) = -\log_e (1.25) = -0.2231$

whence

$M = 43\,800/0.2231 = 196\,000$ hours
If the module contains 500 components, their mean MTBF must be

$500 \times 196\,000 = 98 \times 10^6$ hours

The corresponding mean failure rate is the reciprocal of this, or $1/(98 \times 10^6) = 10.2 \times 10^{-9}$ per hour or 10.2 per 10^9 hours.

This is an extremely low failure rate, but not unattainable; the maintenance records of early American electronic telephone exchanges show component replacement rates of 10 per 10^9 hours for low-power silicon transistors, 5-10 per 10^9 hours for small fixed capacitors and only 1 per 10^9 hours for carbon film and metal film resistors. These exchanges are expected to have a working life of some 25 years so high-reliability components are mostly used to build them.

2.4 Availability

Many instrumentation and control systems are used in situations where repairs cannot be started as soon as a fault occurs. The factor which indicates the probability of successful operation for a certain time is the reliability, which is the most important parameter for the user. This figure is critical, for example for aircraft electronics, where there are no facilities for in-flight maintenance, although when the aircraft returns to the base the faulty unit is normally replaced and sent for subsequent repair. It is also the relevant factor for all satellite electronics which are considered non-repairable, although recent activities from the American Space Shuttle suggest that it may be possible in the near future to make some repairs to satellite equipment.

The manager of, say, an automatic test facility is not interested directly in the MTBF of his or her equipment if he or she can get a fault repaired quickly. The manager's concern is the amount of work that can be put through in a given time; this involves both the time between failures and the repair time. The quantity usually quoted is the availability, defined as the proportion of the switched-on time during which the equipment is available for work.

This can be determined from the running log of the equipment by dividing the total switched-on time into the up-time U during which it is working and the down-time D during which it is faulty or being repaired. The switched-on time is then $U + D$ and the availability is

$$A = U/(U + D) \tag{2.3}$$

The unavailability or down-time ratio is $1 - A$ or $D/(D+U)$.

Where the equipment is used for some sort of batch operation there may be an extra period which must be included in D to allow for restarting the equipment. Thus if automatic test equipment fails part way through a test, that test must be abandoned and after repair it must be repeated. To take this into account, the definition of down-time must include the time to find and repair the fault and the time needed to rerun any aborted test.

In this case there may be a regularly scheduled maintenance period and this is normally excluded from the down-time. However, for continuously running equipment any time devoted to regular maintenance must be classed as down-time.

The availability can also be expressed in terms of the MTBF M and the mean time to repair R as

$$A = M/(M+R) \tag{2.4}$$

This is the asymptotic value to which A converges; it is initially 1 and it decays to within 2% of the final value after a period of about $4R$ from switching on (Shooman 1968).

The assumption of constant failure rate is an acceptable simplification for most purposes, but experience suggests that the mean repair time tends to fall somewhat with time as the maintenance staff acquire experience in diagnosing and repairing faults. However, the staff maintaining some very reliable equipment such as electronic telephone exchanges with a design life of some 25 years encounter faults so rarely that they have little opportunity to gain experience in a working environment. The usual way of providing practice is then to use a simulated environment into which various faults can be deliberately inserted.

2.5 Choosing optimum reliability

When the designer of electronic equipment asks the purchaser what level of reliability he or she requires in the product the initial answer may be 'As high as possible.' In an ideal world this answer may be acceptable but in the real world economic factors generally intrude. A widely used criterion is the total cost of ownership; the design which leads to the lowest total cost is then selected. The cost of ownership includes the initial purchase price, the cost of repairing faults and replacing components and the cost of standby equipment. In some cases, for example telephone exchanges or power generating equipment which is expected to be available 24 hours a day, there is an extra cost involved whenever it is inoperable as some income is lost during the down-time.

As a higher reliability is demanded the costs of design, manufacture and testing all increase rapidly; this is reflected in the purchase price. The costs of repairs and maintenance and the standby equipment, however, all fall as reliability increases, as indicated in Figure 2.3. The total cost of ownership

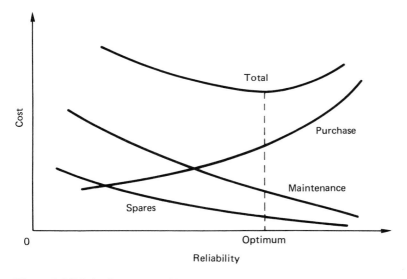

Figure 2.3 *Relation between total life cost and reliability*

over the life of the equipment generally has a minimum value and the design which corresponds to this value is often chosen if no other criteria are involved. For high-reliability systems where failure may imperil human lives there is usually a statutory requirement for a specified minimum reliability and this becomes the overriding factor. The designer's aim is then directed towards achieving this reliability at minimum cost. Such requirements are stipulated for blind-landing control systems for passenger aircraft and for the safety circuits of nuclear power reactors.

The choice of the optimum reliability for some products may be affected by the existence of a guarantee period after purchase during which the manufacturer agrees to repair any fault (not caused by misuse) free of charge. Many domestic products fall into this category, generally having a guarantee period of a year. It is generally reckoned that any service call which has to be made during the guarantee period will largely cancel out the manufacturer's profit. The target MTBF in this case must be considerably greater than the guarantee period (usually 1 year) so that very few of the products fail during this period. Any change which increases product reliability must have its cost compared with the saving in the cost of free service calls which it produces. One change which appears to be advantageous is the substitution of the mechanical timer in a washing machine by an electronic controller with an embedded microprocessor. This gives a marked improvement in reliability at little additional cost and also provides more flexible control.

2.6 Compound systems

In most assessments of reliability we need to evaluate overall system reliability in terms of the reliabilities of the system components.

This depends upon two rules for combining probabilities. The first is the product rule which gives the combined probability that two independent events will both occur as the product of the separate probabilities. Since reliability is the probability of zero failures, if the reliabilities of two components of a system are R_1 and R_2 this gives the system reliability as

$$R = R_1 \times R_2 \tag{2.5}$$

The assumption here is that both components must be working for the system to work.

This result can easily be extended to systems with more than two components as a product of all the component reliabilities. For example, if we consider a telemetry transmitting system consisting of a transducer and signal processing unit, a radio transmitter and aerial and a power unit with reliabilities for a given duty of 0.95, 0.89 and 0.91, the overall system reliability is

$$R = 0.95 \times 0.89 \times 0.91 = 0.77$$

This kind of system in which all units must be working for the system to work is often called a series system. For some purposes we need to examine an alternative situation in which a number of different events may occur and we need to estimate the probability of any one occurring. This is simplest when the events are mutually exclusive, so that only one can occur at a time. This situation is covered by the addition rule which states that if

p_1 and p_2 are the probabilities of the two events occurring separately, the probability of either one occurring is

$$P = p_1 + p_2 \qquad (2.6)$$

Such a situation could be applied to the failure of a diode or transistor for example. If the probability of a short-circuit fault is 9.5 and of an open-circuit fault is 2.4, both per million hours, the probability of either fault occurring is the sum, that is 11.9 per million hours. This situation is one in which the two events are mutually exclusive since a device cannot be simultaneously open circuit and short circuit.

A more frequent case is that in which the two events can both occur. Here we have four possible outcomes:

1 Neither event occurs.
2 Event 1 only occurs.
3 Event 2 only occurs.
4 Both events occur.

The probability of one or more events occurring is the sum of the probabilities of 2, 3 and 4 since these are mutually exclusive. If the probabilities of the two events occurring are p_1 and p_2 the probabilities of these events are:

2 $p_1 \times (1 - p_2)$
3 $p_2 \times (1 - p_1)$
4 $p_1 \times p_2$

Here we use another axiom of probability theory: that is, if the probability of an event occurring is p, the probability of its not occurring is $1 - p$. Thus, in 2, p_2 is the probability that event 2 will occur, so the probability that it will not occur is $1 - p_2$.

Thus the combined probability required is the sum of these:

$$P = p_1 - p_1 p_2 + p_2 - p_1 p_2 + p_1 p_2$$
$$= p_2 + p_2 - p_1 p_2 \qquad (2.7)$$

When estimating the probability of failure of a complex assembly we generally simplify equation (2.7) since the individual probabilities are very small in a system which has an acceptable reliability. Thus the product terms will be negligible and we can estimate the overall probability of failure as the sum of the separate probabilities of all the components.

The probability of system failure is then

$$P = p_1 + p_2 + p_3 + \ldots + p_n \qquad (2.8)$$

We assume that the failure of any components will result in system failure. For example, if the individual component failure rates are of the

20 *Reliability in Instrumentation and Control*

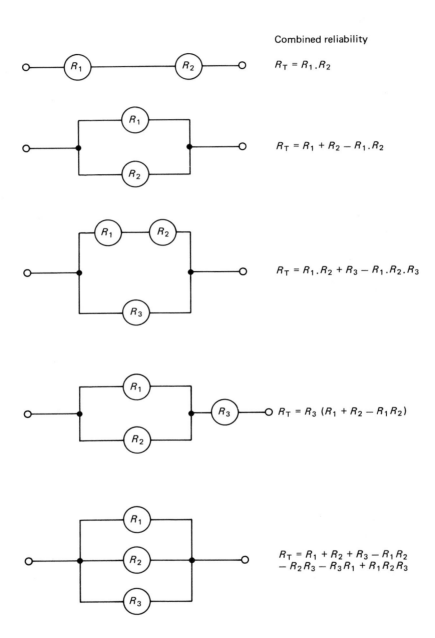

Figure 2.4 *The reliability of compound systems*

order of 10^{-5} over the operating period, the product terms will be of the order of 10^{-10} and will be vastly less than the likely error in the assumed component failure rates. They can thus be safely ignored.

We can apply equation (2.7) to the calculation of the combined reliability of a system comprising two parallel units in which the system operates when one unit or both are working. Here the probabilities p_1 and p_2 are the reliabilities of the two units and the overall reliability is given by

$$R_T = R_1 + R_2 - R_1 R_2 \tag{2.9}$$

For two identical units the combined reliability is

$$R_T = 2R - R^2 \tag{2.10}$$

In a similar manner the combined reliability of a triplicate parallel system in which any one unit can provide service can be shown to be

$$R_T = 3R - 3R^2 + R^3 \tag{2.11}$$

which is often called a 1 out of 3 system.

The probability of failure of one unit is $1 - R$, so that the probability of failure of all three is $(1 - R)^3$. This is the only outcome which gives system failure. The reliability is thus $1 - (1 - R)^3$ which reduces to $3R - 3R^2 + R^3$.

This result assumes that any combining or switching operation required to maintain service has a negligible chance of failure. If this is not the case R_T should be multiplied by the reliability of the switching unit.

Equation (2.11) can be applied to a high-reliability power supply which incorporates three independent units each feeding the common power line through an isolating diode. If the diodes are heavily derated they should have a very low failure rate and the system reliability can be expressed in terms of the reliability of the three units. Using equation (2.11) involves the assumption that any one unit can supply the load on the system.

An alternative method of organizing a triplicate system is to follow it by a majority voting circuit. The system will then operate correctly if any two or all three of the units are working. Assuming identical units, the probability of all three units working is R^3 where the reliability of each unit is R. The probability of two units working and one faulty is $R^2(1 - R)$. This can occur in three ways since each of the three units can be the faulty one. Thus the overall reliability is

$$R_T = R^3 + 3R^2(1 - R) = 3R^2 - 2R^3 \tag{2.12}$$

By combining series and parallel arrangements the reliability of more complex systems can be calculated. Some of these systems and their overall reliability R_T are shown in Figure 2.4.

Bibliography

Ascher, H. and Feingold, H. (1984) *Repairable Systems Reliability*, Marcel Dekker, New York
Barlow, R. E. and Proschan, F. (1965) *Mathematical Theory of Probability*, Wiley, New York
Billington, R. and Allan, R. N. (1983) *Reliability Evaluation of Engineering Systems, Concepts and Techniques*, Pitman, London
Carter, A. D. S. (1972) *Mechanical Reliability*, Wiley, London
Cluley, J. C. (1981) *Electronic Equipment Reliability*, 2nd edn, Macmillan, London
Dhillon, B. S. (1983) *Systems Reliability, Maintainability and Management*, Petrocelli Books, Princeton, NJ
Dhillon, B. S. *et al.* (1980) *Engineering Reliability*, Wiley, New York
Green, A. E. and Bourne, A. J. (1972) *Reliability Technology*, Wiley, London
Henley, E. J. and Kumamoto, H. (1981) *Reliability Engineering and Risk Assessment*, Prentice Hall, Englewood Cliffs, NJ
Jardine, A. K. S. (1973) *Maintenance, Replacement and Reliability*. Pitman, London
Kapur, K. C. (1977) *Reliability in Engineering Design*, Wiley, New York
Klassen, H. B. and van Peppen J. C. L. (1989) *System Reliability*, Edward Arnold, London
Lambert, B. (1990) *How Safe is Safe?* Unwin Hyman, London
Shooman, M. L. (1968) *Probabilistic Reliability: an Engineering Approach*, McGraw-Hill, New York
Smith, D. J. (1973) *Reliability Engineering*, Pitman, New York

3 Reliability assessment

3.1 Component failure rates

An essential part of the design of reliable instrumentation and control systems is a regular assessment of reliability as the design proceeds. The design can then be guided towards the realization of the target MTBF as the finer details are decided and refined where necessary. If reliability assessment is delayed until the design is almost completed there is a danger that the calculated MTBF will be so wide of the target that much design work has to be scrapped and work restarted almost from the beginning.

In the preliminary design phase only an estimation of component numbers will be available and an approximation to the expected MTBF is all that can be computed. As the design proceeds more detail becomes available, more precise component numbers are available and a more accurate calculation is warranted. In the final assessment we can examine the stress under which each component operates and allocate to it a suitable failure rate.

The basis of all reliability assessment of electronic equipment is thus the failure rates of the components from which it is assembled. These rates depend upon the electrical stresses imposed on each component, the environment in which it functions and the period of operation. The effect of the environment and the stress will vary widely, depending upon the type of component involved, but all components tend to show similar trends when failure rates are plotted against time.

3.2 Variation of failure rate with time

The general behaviour of much electronic equipment is shown in Figure 3.1 in which failure rate is plotted against time. In view of its shape this is often called the 'bath-tub' curve. The graph can be divided into three areas: an initial period when the failure rate is falling, usually called 'infant mortality'; a longer period with an approximately constant failure rate corresponding to the normal working life; and a final period with an increasing failure rate, usually called the 'wear-out' phase. This kind of

24 *Reliability in Instrumentation and Control*

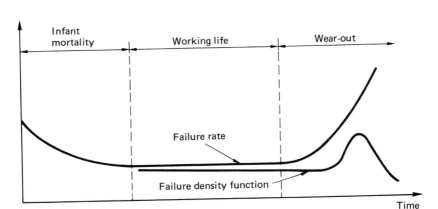

Figure 3.1 *Failure rate and failure density function variation with time*

behaviour was first observed in equipment which used thermionic valves but similar characteristics are also found in transistorized equipment, with a somewhat longer period of infant mortality. The failure rate is subsequently approximately constant. The fault statistics from maintained equipment, however, sometimes show a slowly decreasing failure rate which is usually attributed to an improvement in the quality of the replacement components. The argument is that as time passes the component manufacturer learns from failed components what are the more common causes of failure and can take steps to eliminate them. The manufacturer is thus ascending a 'learning curve' and the reliability of the components is expected to increase slowly with time. The evidence from equipment fault recording and life tests thus either supports the assumption of a constant failure rate, or suggests that it is a slightly pessimistic view.

In a complex item of equipment such as a digital computer there are inevitably a few components with less than average reliability and these are the ones that fail initially. Their replacements are likely to be more reliable, so giving an improvement in system reliability. As these weaker items are eliminated we reach the end of the infant mortality phase and the system failure rate settles down.

Where high reliability is important, for example in aircraft control systems, the customer usually tries to avoid the higher failure rate associated with the infant mortality phase and may specify that the equipment should undergo a 'burn-in' period of operation at the manufacturer's factory before dispatch. The burn-in time is then chosen to cover the expected infant mortality phase (Shooman 1968).

The wear-out phase is never normally reached with modern transistorized equipment as it is almost always scrapped as obsolete long before any sign of wear-out occurs. Even in equipment designed for a long life, typically 25 years, such as submarine repeaters and electronic telephone exchanges, the wear-out phase is not expected to appear.

Any component which has moving parts such as a switch or a relay must, however, experience a wear-out phase which must somehow be avoided where high reliability is needed. Where no repair is possible, for example in satellite control systems, bearings and other points of wear must be of sufficient size and their loading must be suitably low. This will enable the designer to ensure that wear will not degrade performance until the operating period is well above the expected life of the system. Long-life components of this kind need special design and testing and are thus expensive. Where regular maintenance can be undertaken, it may be better to use generally available components and replace them at regular intervals, before any wear causes a decrease in reliability.

3.3 Failure modes

A component can be classed as faulty when its characteristics change sufficiently to prevent the circuit in which it is used from working correctly. The change may be gradual or sudden, classed as degradation or catastrophic failure. Catastrophic failure is due almost invariably to a short-circuit or open-circuit condition; typically diodes and capacitors may develop short circuits and resistors and relay coils may develop open circuits. Generally catastrophic faults are permanent, but some short-circuit faults may be only temporary if they are caused by small conducting pieces of wire which can be moved if the equipment is subject to vibration. Some years ago a major repair effort was needed on a number of avionic units into which a batch of faulty diodes had been built. In a defective manufacturing process some conducting whiskers had been sealed inside a batch of semiconductor diodes. They lay dormant during the normal testing procedures but as soon as the equipment was subjected to severe vibration in service the whiskers migrated and caused intermittent short circuits.

The effect of degradation failure depends upon the type of circuit into which the component is built. Generally analogue circuits are more critical than digital circuits in that a smaller drift in characteristics will cause a fault. In a narrow band filter a change in the value of a capacitor by only 1% may be unacceptable; for example, in an LC tuned circuit resonant at 10 MHz it would shift the resonant frequency by 0.5% or 50 kHz. This would not be acceptable in a system with 25 kHz channel spacing. On the other hand capacitors used for decoupling or power supply filtering may change value to a much greater extent before the circuit will fail. For example, to reduce size and cost electrolytic capacitors are often used to perform these circuit functions and with conservative design the circuits can cope with typical manufacturing tolerances on capacitance of -10% to $+50\%$ without failure. For reliable circuit design it is important to calculate the change which each capacitor can undergo without causing failure and select the most suitable type. Fortunately there are many types of capacitor available from which to choose, each with its own characteristics and tolerances.

The situation is quite different with transistors in that an important characteristic, the common emitter current gain, always has a wide variation. It may be specified for example as within the range 200 to 800. Some manufacturers divide transistors of a given type into grades A, B and C with different ranges of current gain, but even then the range of gain for a given type is quite wide and the designer has to accept this and produce circuits which will work correctly with any expected value of gain. With analogue circuits this usually involves stabilizing the circuit gain by using a substantial amount of negative feedback.

Digital circuits are usually more tolerant of variations in current gain and can thus work satisfactorily with transistors having a wider range of current gain than can most analogue circuits. Thus when considering degradation failure where transistor parameters drift slowly, it may be appropriate to allocate a higher failure rate to transistors used in analogue circuits than to the same transistors used in digital circuits. Catastrophic faults of course will affect both types of circuit equally, causing complete failure. The relative frequencies of degradation and catastrophic faults will determine how much allowance must be made for the type of circuit in which the transistor is used. This is an example of one of the complications attending reliability assessment; the failure rate to be used for a particular component may depend much upon the type of circuit in which it is used.

3.4 The effect of temperature on failure rates

Almost all components exhibit a failure rate which increases with the temperature at which they operate. In many cases, particularly with semiconductor devices, the rate of increase with temperature agrees with that predicted theoretically by a law long used by chemists to relate the speed of a chemical reaction to the temperature at which it occurs. There are two failure modes which can be envisaged in which the speed of a reaction determines the onset of failure; the first is that in which a component has a slightly permeable sealing and is operated in an atmosphere containing contaminants, or despite good sealing a small amount of contaminant is trapped inside the component during manufacture. This may occur even if the component is fabricated in a so-called clean room with well-filtered air. No filter is able to remove all suspended matter and there is always the possibility of a minute amount of contaminant being present which can find its way into a component. In either case the contaminant diffuses into or reacts with the semiconductor and if present in sufficient quantity it will ultimately impair the device characteristics enough to cause failure. The faster the reaction proceeds the sooner will the device fail; consequently if a number of the devices are used in an instrument its failure rate will be proportional to the speed of the reaction (Cluley 1981).

Another failure mechanism is dependent upon the nature of the encapsulation of a component and was a problem with early plastic mouldings. If these were subject to electrical stress and humidity and were in a dirty atmosphere the film of moisture could conduct. The flow of current then damages the surface and leaves a carbonized track which provides a permanent leakage path across the insulator. This fault is more likely to occur if there is some surface defect such as a small crack which provides a preferred leakage path.

There are two expressions used by chemists relating reaction speed and temperature. The simplest of these, which nevertheless agrees well with experimental results, is that due to Arrhenius:

Reaction speed $k = c \times \exp(-b/T)$ (3.1)

where b and c are constants and T is the absolute temperature (Klaasen and van Peppen 1989). This law was originally developed empirically to fit the experimental results and it can be modified by taking logarithms of both sides of the equation to give

$\log k = \log c - b/T$ (3.2)

Thus if the component failure rate λ which is proportional to k is plotted against $1/T$ on log-linear paper the result should be a straight line with a slope of $-b$. If the failure mechanism is known the value of b can be calculated.

This expression is useful to enable the failure rate at one temperature to be calculated, knowing the failure rate at another temperature. It is particularly applied to estimate the failure rate of components at working temperatures from data accumulated in accelerated life tests conducted at high temperatures.

The form of Equation (3.2) shows that the failure rate will change by a fixed ratio for a given small change in T. For reliability assessment it is convenient to use a 10°C increment in T. The corresponding factor for the increase in failure rate varies between about 1.2 and 2 for semiconductors with somewhat larger values for other components. Equation (3.2) can be used to deduce how component failure rates depend upon working temperature; for example, if the factor for failure rate increase is 1.6 for a 10°C rise and tests have established a failure rate of 15 per 10^9 hours for a

28   *Reliability in Instrumentation and Control*

working temperature of 30°C, the effect of increasing the working temperature to 55°C will be to increase the failure rate to 15 × (1.6) 25/10 = 48.6 per 10^9 hours. If accelerated life tests have been conducted at several high temperatures a plot of failure rates against $1/T$ can be used to extrapolate the results down to working temperatures, so saving years of testing time (Rouhof 1975).

3.5 Estimating component temperature

We have seen that component failure rates are markedly dependent upon temperature; it is thus important in any reliability assessment to establish the temperature limits that various components experience. This is relatively easy for components such as low-loss capacitors which do not dissipate any appreciable amount of heat and thus assume the same temperature as their surroundings. However, any component which absorbs electrical energy will dissipate this as heat and will thus be at a higher temperature than its surroundings. The maximum internal temperature for reliable operation is normally specified by the manufacturer and so as the ambient temperature increases the power dissipated must be reduced. This process is called 'derating'. Typical circuit design standards for carbon and metal film resistors suggest using the recommended power rating for ambient temperatures up to 45°C and derating linearly down to zero dissipation at 90°C. This is for normal free-air conditions; if forced cooling is used higher dissipation can be allowed.

Some degree of derating is suggested for electronic equipment in commercial aircraft (Arsenault and Roberts 1980) as follows:

Tantalum capacitors	$T_C = T_M \times 0.67$
Aluminium electrolytic, paper, ceramic, glass and mica capacitors	$T_C = T_M \times 0.72$
All resistors	$T_C = T_M \times 0.68$ or 120°C whichever is lower
Semiconductors, ICs, diodes	Max. junction temp. = $T_{MJ} \times 0.6$
Relays and switches	$T_C = T_M \times 0.75$
Transformers, coils and chokes	T_C to be at least 35°C below the manufacturer's hot spot temperature

Here T_C is the external surface or case temperature of a component, T_M the component manufacturer's maximum permissible body temperature at zero dissipation, and T_{MJ} the maximum junction temperature specified by the manufacturer with zero dissipation.

Where extremely high reliability is required the derating factors 0.67, 0.72, etc., may be reduced but this will involve using larger resistors and heat sinks so increasing equipment volume.

A critical factor in determining transistor reliability is the junction temperature; consequently it is important that the circuit designer should be able to estimate this fairly accurately. The usual basis for this is the analogy between the flow of heat and the flow of electric current. If we have a current source of strength I A at a potential of V_1 connected via a resistance of R to a sink at a potential of V_2 the relation between the potentials is

$$V_2 = V_1 + I \times R$$

The analogues of potential and current are temperature and thermal power (usually expressed in watts) so giving the equation

$$T_2 = T_1 + \theta \times W \tag{3.3}$$

Here temperatures are in °C. θ is thermal resistance in units of kelvin per watt and W is the power generated by the source in watts. In the case of a transistor T_2 is the junction temperature and T_1 the ambient temperature. The thermal resistance has two components: θ_I the internal resistance between the junction and the transistor case and θ_E the external resistance between the transistor case and the surroundings. θ_I is fixed by the design of the transistor and cannot be changed by the user, but θ_E depends upon the transistor mounting. In order to minimize the junction temperature θ_E can be decreased by mounting the transistor on a heat sink, a block of metal which is a good heat conductor (usually aluminium) with fins to help to dissipate heat and painted black.

For a BFY50, a metal-cased low-power transistor rated at 800 mW maximum power, the value of θ_I is 35 K per watt and with no heat sink θ_E is 185 K per watt. If we have a transistor dissipation of 800 mW and an ambient temperature of 30°C the junction temperature without heat sink would be

$$T_2 = 30 + 0.8(35 + 185) = 195°C$$

This is much too high for reliable operation, so a heat sink will be needed to improve the cooling. If we specify that the junction temperature must not exceed 100°C we can use Equation (3.3) to calculate the thermal resistance of a suitable heat sink. Thus

$$T_2 = 100 = 30 + 0.8\,(35 + \theta_E)$$

whence $\theta_E = 52.5$ K per watt. A corrugated press-on heat sink is available for transistors similar to the BFY50 (TO39 outline) with a thermal resistance of 48 K per watt which would provide adequate natural cooling. If the dissipation or the ambient temperature were higher forced air cooling would be needed.

The above calculation assumes that the transistor dissipation is constant. If the transistor is pulsed the situation depends upon the relation between the pulse duration and its thermal time constant. If the pulse duration is much smaller than the time constant the important factor is the average dissipation which can be used in Equation (3.3) to determine the junction temperature. This will not fluctuate appreciably. However, if the pulse duration becomes comparable with the time constant it will be necessary to investigate the heating and cooling of the junction and determine its maximum temperature.

3.6 The effect of operating voltage on failure rates

The main effect of a change in the working voltage on resistor reliability is that caused by the change in dissipation and so in the working temperature. Where resistors are used in low-power applications so that their temperature rise above ambient is small the operating voltage will not significantly affect their failure rate.

There is some evidence that using voltage derating reduces the failure rate of semiconductors and some users ensure that the manufacturer's maximum rated voltage is at least twice the operating voltage in order to improve reliability. This procedure can be applied to discrete transistors and diodes but not to most digital devices. Packages such as transistor–transistor and emitter-coupled logic and most storage devices are designed to operate at fixed voltages, with tolerances of typically ±5%, so that any attempt to improve reliability by reducing the supply voltage will prevent circuit operation. CMOS digital circuits on the other hand are typically specified for operation with a supply voltage of 3 to 15 volts. This allows scope for some voltage derating but there is a disadvantage in that the speed of operation is roughly proportional to the supply voltage. Thus voltage derating will result in longer switching times and slower operation.

Analogue integrated circuits such as operational amplifiers on the other hand can operate with a range of supply voltage, typically ±5 to ±18 V or ±3 to ±18 V. Some degree of voltage derating can be applied, but the maximum output voltage swing is approximately proportional to the supply voltage so that the consequence of derating is a reduction in the maximum output voltage (Sinnadurai and Roberts 1983).

Experience with paper dielectric capacitors showed a marked dependence of failure rates on operating voltage and an empirical law which fitted the results showed a fifth power relation with the failure rate being proportional to $(V/V_R)^5$. Here V is the peak operating voltage and V_R the rated voltage. The same relation can also be applied to polyester capacitors which are widely used in sizes up to about 1 μF (Arsenault and Roberts 1980). These are made with DC voltage ratings from 63 to 400 V so that substantial voltage derating can be used in typical transistor circuits with supply potentials of 20 V or less.

For example, if a polyester capacitor rated at 40 V AC has a failure rate of 12.5 per 10^9 hours at its working temperature, derating it to 25 V AC should give a failure rate of
$12.5 \times (25/40)^5 = 1.19$ per 10^9 hours

3.7 Accelerated life tests

The failure rates of modern components are so low at typical operating temperatures that any life tests under these circumstances will involve many thousands of components on test for long periods. For example, if we wish to test components which have an expected failure rate of 20 per 10^9 hours we need to put, say, 5000 on test for just over a year before a failure is likely. We need more than one fault to produce useful data so that a test such as this is bound to be very lengthy. We can obtain useful data in a much shorter time by artificially increasing the failure rate by a known amount, generally by operating the components at a high temperature. Transistors for submarine cable repeaters are required to have a working life of some 20 years and are tested typically at junction temperatures of 250°C and 280°C (Rouhof 1975). Integrated circuits for electronic telephone exchanges are also required to have a 20-year life, but the reliability is not quite so important as the cost of repairing a fault in an office or an exchange is much less than in a submarine repeater. Typical test temperatures for these are 125°C, 150°C and 160°C. The acceleration factor for the test at 160°C compared with a working temperature of 70°C is calculated as 555 so that a test of 320 hours' duration represents a working life of 20 years (Sinnadurai and Roberts 1983). If we make a somewhat gross approximation that a 10°C rise in temperature causes an increase of failure rate by a fixed ratio, the ratio in this case is about 2.02.

Although accelerated life tests of semiconductors at elevated temperatures are widely used they are sometimes deceptive as the failure mechanism changes at high temperatures and the straight-line extrapolation down to working temperatures fails as the graph changes slope. Despite this the high-temperature tests are valuable for comparing one component design with another, and also for producing faults in a reasonably short time for failure mode analysis.

3.8 Component screening

Since the failure rate of nearly all components shows an initial value greater than that measured after the 'infant mortality' period has elapsed it would improve system reliability if these initial failures could be weeded out. Some can be eliminated by a few hundred hours of 'soak test' at the maximum working temperature but this will not find all early semiconductor failures. It is thus common to give these separate screening tests at typically 125°C for 168 hours. The components are tested at ambient temperature, usually 25°C, before and after the screening and any item which shows a significant change in characteristics is rejected. For transistors the most sensitive parameter is common emitter current gain; for operational amplifiers open-loop gain and cut-off frequency f_T. In addition to electrical tests, thermal cycling between temperature limits, vibration testing and acceleration testing in a centrifuge may be used for screening out mechanically weak items (Jensen et al. 1982). The bonding of the leads in current semiconductors is so effective that accelerations of $30\,000\,g$ are used for screening (Arsenault and Roberts 1980).

Data collected by AT&T from telephone exchanges (Holcomb and North 1985) suggests that the infant mortality phase can last for up to a year for equipment used in such benign surroundings. Thus high-temperature screening with an acceleration factor of, say, 100 will require some 87 hours to eliminate all likely early failures.

3.9 Confidence limits and confidence level

A problem which arises in evaluating system reliability from the failure rates of its components is the relation between the figures to be used in design calculations and the results of life tests and equipment fault records. All life tests are carried out on only a sample of the components manufactured and their validity depends upon the degree to which the sample is representative of the full population. The sample should be selected randomly but even so its characteristics will not be exactly the same as those of the whole population, and will vary from sample to sample. Thus any data from life tests cannot be applied directly to the general population, but we can assign a probability called the confidence level that its failure rate lies between prescribed limits, called the confidence limits. Thus failure rates are stated in the form 'The failure rate of this component lies between 3.6×10^{-9} and 7.5×10^{-9} per hour with a confidence level of 90%.' This implies that the probability that the failure rate lies between the limits quoted is 90%. Figures given in BS 4200: Part 7: 1982 show the number of component hours of testing needed to establish a specific failure rate with a 60% confidence level. Here we are interested only in the upper confidence limit, giving us an upper bound for the failure rate, so the lower limit is ignored. For a demonstrated failure rate of 10^{-8} per hour (10 FITs) the testing times are in millions of component hours:

No faults   91.7
1 fault   203
2 faults   311
3 faults   418
4 faults   524
5 faults   630

This is a considerably shorter test than would be required for a confidence level of 90%, in which for zero faults the test time would be 230 million hours and 389 million for one fault.

3.10 Assembly screening

In addition to component screening, circuit boards are often subjected to screening. This is unnecessary if the complete system is screened, but any boards destined for spares must be subjected to the same treatment as the system. The usual tests may include temperature cycling, a soak test and vibration. For American military and commercial equipment a range of test programmes are used. For equipment used in an air-conditioned room with little temperature variation, temperature limits of 20 to 30°C are specified, whereas equipment used in a fighter aircraft is subject to rapid and wide-ranging temperature variations and would be cycled between -54 and $71°C$ whilst undergoing vibration and with the power supply periodically switched on and off. The test conditions are designed to represent the most arduous environment that the equipment is expected to encounter in service.

3.11 Dealing with the wear-out phase

The usual method of estimating system failure rate is to add together the individual failure rates of all the components in the system. This assumes that we are dealing with a minimum system with no redundant items, and that every component must be in working order if the system is to work properly. This simplified calculation ignores the probability of multiple faults, but modern components have such a low failure rate that this is a justified assumption for any normal working period.

In addition it is assumed that component failure rates are constant over the period considered. This means that none of them will have reached the 'wear-out' phase of their working life. In repairable systems it may be necessary to replace some components with moving parts at regular intervals to ensure that they do not reach the 'wear-out' phase. Otherwise, as their working life is less than that of most electronic components, they may severely limit the system reliability.

This procedure may be applied to small switches which may be expected to survive a million operations before failure. If such a switch is operated 100 times per day it should survive for just over 27 years. If, however, it is used in a hostile environment which reduces its expected life by a factor of 10, or it is operated 1000 times per day, the survival time is reduced to only 2.7 years which may be much less than the expected operating period. It may then be necessary to replace the switches at, say, yearly intervals to ensure that they do not become a significant factor in reducing system reliability.

If test data on the time to failure of a batch of components is available a more precise estimate of the appropriate replacement policy can be made. This is based upon the failure rate or hazard rate curve shown in Figure 3.1 and a curve derived from it called the failure density function, also shown in Figure 3.1. The failure rate is defined as the ratio of the number of failures Δn occurring in the interval from t to $t + \Delta t$ to the number of survivors $n(t)$ at time t, divided by the interval Δt (Shooman 1968, Klaasen and van Peppern 1989). It is thus

$$z(t) = \Delta n/(\Delta t \times n(t))$$

The failure density function $f(t)$ is defined as the ratio of the number of failures Δn occurring in the interval from t to $t + \Delta t$ to the size of the original test batch $n(0)$ at time $t = 0$. Whence

$$f(t) = \Delta n/(\Delta t \times n(0))$$

In many situations we expect very few failures during the operating period of reliable equipment, so $z(t)$ will be almost identical, but if we continue well into the wear-out phase during life tests we expect most of the

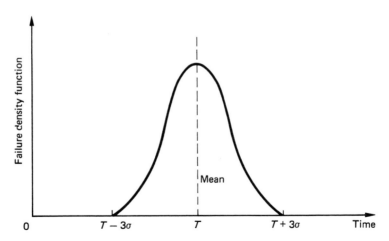

Figure 3.2 *Normally distributed failure density function*

items to fail. Then $f(t)$ and $z(t)$ will diverge as shown in Figure 3.1. If possible we need to acquire enough data to plot approximately the wear-out peak in the failure density function $f(t)$. This is often a bell-shaped curve, as shown in Figure 3.2, the 'normal' distribution. This arises when the variations in the time to failure are due to a number of causes, each producing a small change. The data used to plot the curve can be used to estimate the mean life T and the standard deviation σ. Tables of the distribution show that only 0.13% of items will have a time to failure less than $T - 3\sigma$, so this is often taken as the working life of the items (Caplen 1972). In a system expected to operate for a longer period, the items should be replaced at this time.

3.12 Estimating system failure rate

We take as an example for estimating unit reliability a small assembly consisting of the components in the table below:

Component	Failure rate per 10^9 hours	No.	Total failure rate per 10^9 hours
Tantalum capacitor	3.0	2	6.0
Ceramic capacitor	2.8	8	22.4
Resistor	0.42	11	4.62
Transistor	8.3	3	24.9
Integrated circuit	12.4	2	24.8
Diodes	4.0	5	20.0
			102.7

The MTBF is thus $10^9/102.7 = 9.74 \times 10^6$ hours.

If five of these assemblies are used in a non-redundant satellite control system which is expected to have a 7-year life the total failure rate is $5 \times 102.7 = 513.5$ per 10^9 hours.

The MTBF is thus $10^9/513.5 = = 1.95 \times 10^6$ hours and the reliability is

$R = \exp(-T/M) = \exp(-7 \times 8760/1.95 \times 10^6)$
$= 0.969$

This calculation assumes that each assembly has no redundancy and it will thus not work satisfactorily if any component fails.

The expected number of faults in the 7-year operating period is $513.5 \times 7 \times 8760/10^9 = 0.0315$. Since $1/0.0315 = 31.8$ we may expect that if we had 32 satellites in orbit we would expect one of them to fail during its planned life.

3.13 Parallel systems

We have seen that in series systems the overall reliability is the product of the reliabilities of the units which comprise the system. Thus the overall reliability must be less than that of any of the units. Using a parallel configuration we can construct systems which have a greater reliability than that of any of the units. This arrangement is thus of interest to designers of high-reliability systems. The system is usually designed so that any one of the parallel units can provide satisfactory service and some switching mechanism is used to isolate a faulty unit. A simple example is a triple power supply for a critical control system in which the load can be supplied either by a regulated power supply connected to the mains, a float-charged battery or a standby diesel generator. Each source is connected to the load through an isolating diode. If the reliabilities of the three sources are $R_1 = 0.92$, $R_2 = 0.95$ and $R_3 = 0.90$ for a particular operating period, the overall reliability will be: $R_T = R_1 + R_2 + R_3 - R_1R_2 - R_2R_3 - R_3R_1 + R_1R_2R_3 = 2.77 - 0.874 - 0.855 - 0.828 + 0.7866 = 0.9996$.

The expression for R_T is derived easily by considering the probability that the system will not work. This can happen only if all three units are faulty, the probability of this event being $R_F = (1 - R_1)(1 - R_2)(1 - R_3)$. The overall reliability (the probability of the system working) is the complement of this or $(1 - R_F)$, which gives the expression for R_T shown above. In this calculation we have assumed that the switching action of the isolating diodes cannot fail. This is clearly an optimistic assumption, although in some cases the diodes are duplicated to improve the reliability of the switching action. If they are also derated their failure probability will be very much less than the unit failure probabilities implicit in the above figures, and so it can justifiably be neglected. On the other hand if the diode failure rate is likely to be a significant factor it can be incorporated in the unit reliability figures. Thus if the diode reliability is estimated as R_D, the reliabilities of the three separate power sources must be changed to $0.92 \times R_D$, $0.95 \times R_D$ and $0.90 \times R_D$.

In some situations we may have a combination of series and parallel units; for example, in an instrumentation system involving a telemetry link from a moving vehicle there may be a power supply unit, a signal converter and multiplexor unit and twin radio transmitters and aerials. The aerials may be separated to allow one to illuminate the shadows caused by buildings in the field of the other. Generally the output of one aerial alone will give a useful service, so we can evaluate the overall reliability by assuming that the system gives an acceptable performance with only one transmitter operating. The reliability diagram is then as shown in Figure 3.3.

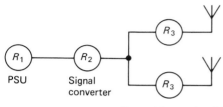

Figure 3.3 *Reliability diagram of telemetry transmitting system*

This can be analysed as a series system of three items, the PSU, the signal converter and the transmitter/aerial combination. The respective reliabilities are: R_1, R_2, and $2R_3 - R_3^2$. The overall reliability is thus $R_1 \times R_2(2R_3 - R_3^2)$.

Other compound systems can be analysed by using the expressions given in Figure 2.4.

3.14 Environmental testing

Although estimates of MTBF are needed during the design and development phase of electronic systems, users often expect some demonstration of the claimed figures. This is particularly the case with military and avionic systems where the specification generally calls for a minimum demonstrated MTBF.

This requires samples of the production run to be tested under the most arduous working conditions likely to be encountered in service. These may vary widely, depending upon the working environment. Thus instrumentation systems to be installed in power station control rooms will be working in a benign environment with no vibration, comparatively small temperature and humidity variations, a stable mains supply and free from dust and water spray. They can thus be tested adequately in a normal laboratory without extra facilities. The transducers and their associated electronics, however, which supply data to the rest of the system, are attached to boilers, turbines, alternators, etc., and so will be subjected to some vibration and high temperatures. These conditions must be incorporated into any realistic test programme. The vibration may be a random waveform or a sinusoidal signal swept over a prescribed range of frequencies, and some specifications may also call for a number of shocks to be applied. Even more severe conditions are experienced by flight test equipment which will be subject to considerable shock and vibration, with rapid and large changes in temperature and rapid changes in humidity. It is also likely to encounter changes in power supply voltage and frequency considerably larger than those typical of a ground environment. All of these factors must be represented in a full environmental test which requires the use of an environmental test chamber. This allows the user to regulate all of the conditions mentioned above, and also to subject the equipment under test to fresh- and salt-water spray and simulated solar radiation. Its electrical performance should also be monitored during the test. Where the equipment may be subjected to electromagnetic fields provision must be made to generate such fields of known strength and frequency during the test to simulate service conditions. Also, where the equipment is to be used near sensitive radio apparatus, there may be a specified limit to the field strength it generates. To check this property a field strength measuring set which covers the appropriate frequency is required. Where the power supply may be contaminated by short high-voltage transients such as those caused when switching inductive loads, provision must be made to inject transients into the power supplied to the equipment on test and to check that no malfunctions are caused by them.

Control equipment in satellites is subject to ionizing radiation not experienced by ground-based apparatus and its resistance to this radiation must be checked in detailed environmental tests.

Generally the testing of equipment to be used in benign surroundings such as computer rooms and laboratories requires a minimum of environmental control, but as the environment to be simulated becomes more severe, with greater excursions of temperature, humidity, etc., and increasing vibration levels, the complexity and cost of simulating the working environment becomes much greater.

Bibliography

Arsenault, J. E. and Roberts, J. A. (1980) *Reliability and Maintainability of Electronic Systems*, Computer Science Press, New York

Billinton, R. and Allen, R. (1983) *Reliability Evaluation of Engineering Systems, Concepts and Techniques*, Pitman, London

Blanks, H. S. (1980) The temperature dependence of component failure rates. *Microelectronics and Reliability*, **20**, 297

BS 4200: Part 7: 1982, British Standards Institution, Milton Keynes

Caplen, R. (1972) *A Practical Approach to Reliability*, Business Books, London

Cluley, J. C. (1981) *Electronic Equipment Reliability*, 2nd edn, Macmillan, London

Dhillon, B. S. et al. (1980) *Engineering Reliability*, Wiley, New York

Jensen, F. et al. (1982) *Burn-in: An engineering approach to the design and analysis of burn-in procedures*, Wiley, Chichester

Klaasen, H. B. and van Peppen, J. (1989) *System Reliability*, Edward Arnold, London

Holcomb, D. P. and North, J.C. (1985) AT&T Tech Journal, **64**, 15

Miller, R. and Nelson, W. (1983) Optimum simple step-stress plans for accelerated life testing. *IEEE Transactions on Reliability*, **R-32**, No. 1

Rickers, H. C. and Manno, P. F. (1980) Microprocessor and LSI microcircuit reliability prediction model. *IEEE Transactions on Reliability*, **R-29**, No. 3

Rouhof, H. W. (1975) Accelerated life test results on submarine cable transistors. *IEEE Transactions on Reliability*, **R-24**, No. 4

Shooman, M.L. (1968) *Probabilistic Reliability: An engineering approch*, McGraw-Hill, New York

Sinnadurai, N. and Roberts, D. (1983) Assessment of micropackaged integrated circuits in high reliability applications. *Microelectronics Journal*, **14, No. 2**

4 System design

4.1 Signal coding

All but the most elementary control and instrumentation systems involve several units such as transducers, amplifiers, recorders and actuators which are connected together to perform the task required. An important feature of the system is the manner in which the signals passed between units are coded.

All early systems used analogue coding in which the signal which represents a system variable has a magnitude proportional to it. For example, an oil lamp invented by Philon in about 250 BC used a float regulator in which the movement of the regulator was proportional to the level of oil.

One of the earliest applications of an analogue controller for industrial purposes was the rotating ball governor used to control the speed of steam engines, developed in 1769 by James Watt. The geometry ensures that whenever the engine speed increases above the desired value the steam supply is reduced, so ensuring that any changes in speed are kept to a minimum.

Analogue techniques continued to dominate control and instrumentation systems until the early 1950s, all the feedback schemes developed during the war for the precise positioning of anti-aircraft guns, radar aerials, etc., being analogue systems.

Analogue coding is simple and in many ways convenient since most transducers deliver analogue outputs and many analogue indicators such as pointer meters and chart recorders are simple and relatively cheap. However, it is difficult to store analogue signals for later display and processing, and to transmit them over long distances without introducing noise and errors.

Many analogue systems use electrical voltage as the quantity which represents the physical variables since this is the output coding of most transducers and a wide range of voltage measuring devices are available to display output quantities.

The accuracy of the display is dependent partly upon the degree to which the voltage across the display device equals the output of the signal processing element. In circuit terms, this means a negligible voltage drop in the wiring between them. Such a condition can easily be satisfied in a compact system, but installations which cover a large area such as power station instrumentation schemes may include some wiring runs which are many hundreds of metres long. In these circumstances the voltage drop in the wiring can no longer be neglected. However, if we change the analogue quantity used from voltage to current the voltage drop, if not too large, will not affect the current in the line and we can preserve the system accuracy. Accurate current generators are not difficult to design and this arrangement is widely used in many boiler house and process control systems which involve long runs of cable. Although the use of current as the analogue quantity almost eliminates the effect of conductor resistance on display accuracy, it requires the maintenance of a high insulation resistance between the pair of wires feeding a display as any current leakage between them will degrade the accuracy of the display. In the hostile environment of a typical boiler house there may be heat and damp to contend with and the avoidance of leakage requires great care with both the installation and the maintenance of all the system wiring.

The advent in the 1950s of transistors which were fast enough to be used in small digital computers enabled such machines to be made smaller, lighter, cheaper and much more reliable than the valve machines which preceded them. They could consequently be used as the nucleus of a control or instrumentation system without incurring any major cost penalty. Since they could also complete complex calculations in a very short time and could easily store masses of data for as long as needed they soon became the natural choice for major control and instrumentation systems.

When microcomputers were developed in the early 1970s it became possible to fabricate the central components of a small digital computer on a few silicon chips, each one only a few millimetres square. The manufacturing process was a high-volume operation and so the cost and size of the computing element fell by a factor of some hundreds. This immensely widened the scope of digital control and measuring systems since they could be used in much cheaper products without incurring extra cost. For example, they were built into the weighing machines used in retail shops, petrol pumps and domestic appliances such as washing machines and microwave ovens. By this time microprocessors had been developed to include storage for both program and data, interface registers, counter/timers and DACs and ADCs. Thus a working system needed only a few packages in addition to the microprocessor itself.

4.2 Digitally coded systems

The availability of fast arithmetic operations enables much signal processing to be performed in real time so that results can be displayed quickly and, generally more important, can be used in a feedback control system. Examples of typical calculations are fast Fourier analysis and the RMS value of a sampled waveform. Even simple 8-bit microprocessors such as the Motorola 6809 can multiply two 8-bit numbers together in only microseconds.

A further advantage of digital coding is its ability to withstand signal attenuation and noise. For instance, many digital systems use signal levels consistent with those of TTL circuits. In the original versions of these circuits, binary 1 is coded as a voltage between 2.4 and 3.5 when transmitted. The receiving device will accept this as a 1 if it exceeds 2.0 V. The interconnection can thus tolerate a negative induced noise transient of 0.4 V for a 1 signal. The maximum transmitted level for a logic 0 is 0.4 V and an input will be recognized as a 0 if it does not exceed 0.8 V. Thus in this logic state the circuit will also tolerate a positive induced transient of 0.4 V (the noise margin) before an error is introduced.

Later versions of TTL with Schottky diodes incorporated to reduce transistor saturation had a somewhat larger noise margin in the 1 state but the same in the 0 state.

If we compare this with an analogue system with a full-scale signal of 10 V, the same induced voltage of 0.4 V will give an error of 0.4/10 or 4%. If the full-scale voltage is 5 V the error will be 8%. TTL packages are used where high currents are needed, for example to drive the solenoids in a printer, but much less current is needed to drive other logic packages and MOS logic suffices for this. These circuits will operate on any power supply voltage from 3 to 15 V, but take longer to switch than TTL. As they have complementary transistors in the output stage their noise margin is about 40% of the supply voltage, that is 2 V for a 5 V supply and 4 V for a 10 V supply. With a 10 volt analogue system a 4 volt noise component would give an error of 40% which would be quite unacceptable. Although these figures show that digital systems can tolerate much more induced noise than analogue systems, the comparison is not quite as marked as indicated since the analogue system will usually require a much smaller signal bandwidth than digital systems. Thus short noise pulses which would cause an error in a digital system can be much reduced in analogue systems by passing the signals through a low-pass filter which can be designed to pass the relatively low-frequency analogue signals but provide considerable attenuation to the noise pulses. However, low-frequency noise such as mains interference will not be affected by such filtering.

A further advantage of digital coding is that by adding extra digits to a byte, or a longer packet of data, error detection or error correction can be provided. The simpler schemes can cope only with a single error, so the size of a block to which the redundant check bits are added must be small enough to make the probability of two errors insignificant.

4.3 Performance margins in system design

Although when describing how a control or instrumentation system operates it is customary to start with the system inputs and follow their path to the output devices, the reverse process is needed for system design. Thus if we are designing a ship's automatic steering control the first data needed is the maximum torque which must be exerted on the rudder to maintain a given course and the maximum rate at which the rudder angle must be altered. These two items enable the power level of the steering motor and its gearing to be calculated, as well as the electrical input power needed by the motor. This must be controllable in magnitude and polarity by low-voltage circuits, often incorporating a microprocessor, which sense the ship's actual heading and compare it with the requested course. A large degree of power amplification is needed to supply the steering motor since the control input may be only 25 mW (5 V and 5 mA) and the motor may require many kilowatts.

An important factor in designing a reliable system is the provision of adequate performance margins in its various components. In this case reserve motor power should be provided to allow for such factors as wear, the increasing frictional load caused by corrosion and the extra rudder angular velocity needed for emergency manoeuvres. An equal reserve of both output power and gain must be provided in the power amplifier which supplies the motor.

Safety margins of this kind are mainly decided by engineering judgement and previous experience since there is no exact method of deciding their size. However, where we need to allow for known tolerances on various system parameters we can make a more exact estimate.

4.4 Coping with tolerances

In order to illustrate the way in which parameter tolerances can be handled we take as an example an instrumentation system with transducers, signal amplifiers, signal processors such as phase-sensitive detectors and some indicating or recording device.

Each of these will have a characteristic such as gain or in the case of a transducer the output expected for a given stimulus. In the case of a force transducer this might be millivolt output per kilogram of force exerted on it. This is specified by the manufacturer with an associated tolerance. When the system is first installed its overall performance will normally be calibrated, in this case by placing a known weight on the transducer and adjusting the system gain until the output device reads correctly.

The calibration test will normally be repeated at regular intervals, their spacing being determined by the rate at which the performance drifts and the maximum permissible system error.

When considering system design it is important to allow an adequate performance margin overall to allow the system to be restored to its specified performance during the whole of its expected life. This means that the maximum expected variation in all system characteristics must be estimated to allow the likely variations in the end-to-end characteristic to be assessed. The usual procedure is the 'worst-case' design in which all parameters are assigned values, usually at the extreme bounds of their tolerance, which will cause the greatest effect on the overall characteristic. The variation in the overall performance can then be calculated and provision made to cope with it. If we start with all parameters at their minimum value we can determine the least value of gain needed in the signal amplifier to enable the system to be calibrated. We assume that a number of systems are being constructed, all to the same design. We must then consider a system with all the parameters at their maximum values, and calculate its performance. This will be well in excess of the specified performance and we can determine how much adjustable attenuation must be included in the system to enable it to be correctly calibrated.

As a simple example we will consider a weighing system consisting of a strain gauge transducer (sensitivity $2\,\text{mV} \pm 5\%$ per kg) followed by an amplifier/phase-sensitive detector (PSD) output $G \pm 10\%$ V DC for 1 V RMS input) which drives a digital display requiring $10\,\text{V} \pm 3\%$ for the full-scale indication of 10 kg. The arrangement is shown in Figure 4.1.

The minimum transducer output for the full-scale load of 10 kg is $20\,\text{mV} - 5\% = 19\,\text{mV}$. A display having minimum sensitivity will require an input of $10\,\text{V} + 3\% = 10.3\,\text{V}$.

Thus the minimum amplifier gain needed is $10.3/(19 \times 10^{-3}) = 542$. This must be the minimum gain, i.e. $G - 10\%$, whence the nominal gain $G = 542/0.9 = 602$. The maximum gain is 10% above this, i.e. $602 \times 1.1 = 662$.

The maximum transducer output is $20\,\text{mV} + 5\% = 21\,\text{mV}$ for full-scale load. This will produce a display input of $662 \times 0.021\,\text{V} = 13.9\,\text{V}$.

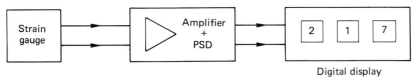

Figure 4.1 Strain gauge instrumentation

The display will require 10 V − 3% = 9.7 V at maximum sensitivity. Thus the attenuation ratio needed to enable the system to be calibrated in all circumstances is 13.9/9.7 = 1.43. If the input resistance of the display is large compared with R_1 and R_2 this gives the condition $(R_1 + R_2)/R_2 > 1.43$.

4.5 Component tolerances

In order to design reliable electronic equipment it is essential to allow for changes in device characteristics during their working life as well as the tolerances in their initial values. Generally tolerances are specified as measured just before components are delivered to the user. There is usually some delay before the components are assembled, so tolerances at assembly will be a little wider to allow for small changes during storage. For example, carbon film resistors typically have a maximum drift during storage of 2% per year whereas metal film resistors which are much more stable are quoted as having a maximum drift of only 0.1% per year. The assembly process usually involves soldering the component to a printed circuit board and so heating it momentarily. This is also likely to cause a small change in the component value. Some equipment will be screened, which usually involves a short period of operation at high temperature and may cause a small change in value. Finally the value will drift during the working life of the equipment. The end-of-life tolerance which the designer must allow for

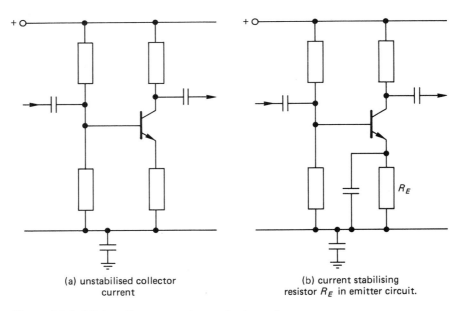

(a) unstabilised collector current

(b) current stabilising resistor R_E in emitter circuit.

Figure 4.2 *Stabilizing collector current by use of emitter resistor*

is thus significantly greater than that measured immediately after the component is manufactured.

Both passive and semiconductor components will experience a drift in characteristics, but as semiconductor tolerances in parameters such as current gain are so large the comparatively small changes during assembly and normal life pose little problem to the designer. For example, if a transistor is specified to have a current gain in the range 100-300, any circuit which can accept this wide variation can easily cope with a 5% drift during assembly and service.

A circuit block often required in instrumentation and control systems is a voltage amplifier having a closely specified gain. In view of the wide variation in open-loop gain caused by transistor tolerances, the customary way of meeting the requirement is to use overall negative feedback. As this is increased the closed-loop gain depends increasingly upon the attenuation in the passive feedback path. At its simplest this will be the ratio of two fixed resistors. Thus we can cope with wide variations in the characteristics of the active amplifier components if we can ensure constant resistance values in the feedback path. This is a much easier task, since we can obtain metal oxide resistors which at low power levels will drift less than 0.1% (film temperature 30°C) during a 25-year life (Osbourne 1980). This application is for submarine repeaters, but a similar requirement for long life in an environment which precludes repair arises in the control systems of commercial satellites. The same resistor is estimated to have a drift of just over 1% in 25 years at 70°C.

Many low-power amplifier requirements are conveniently met using integrated circuits. These generally have even wider gain tolerances than discrete transistor amplifiers and only the minimum gain is usually specified. They are usually operated with a high degree of feedback to stabilize the overall gain and reduce distortion; again the performance is dependent upon resistor stability.

4.6 Temperature effects

Some environments in which control and instrumentation systems operate, such as manned control rooms, have a measure of temperature regulation and the equipment they house is subject to only small temperature variations. At the other extreme electronic engine controllers used in aircraft are mounted near to jet engines and may thus suffer wide temperature variations. For reliable performance the equipment designer must investigate the component changes caused by temperature variations and ensure that they will not prevent system operation.

For example, metal film resistors typically have a temperature coefficient of +50 parts per million (ppm). A temperature change of 80°C will cause a change in resistance of only 0.4% which is less than the manufacturer's tolerance of ±1% and should not prevent most circuits from operating correctly. In many cases, particularly feedback amplifiers, the important factor is the ratio of two resistors rather than their absolute value. If the same type of resistor is used both resistors should change by nearly the same proportion and their ratio will change very little. Where high precision is important wire-wound resistors having a very low temperature coefficient of around +5 ppm are available, but owing to their inductance they are not suitable for use at high frequencies.

The most stable capacitors for values up to 10 nF are silvered mica types which have a typical temperature coefficient of +35 ppm, so that for most purposes the change due to temperature variations can be neglected.

Inductors also have significant temperature coefficients which can be minimized by using a single-layer air-cored coil. This results in coefficients of 5 to 15 ppm. Low-inductance coils wound on ceramic formers, or better with the low-expansion conductor deposited in a groove on the surface of the former, yield coefficients of around 1 ppm.

These low temperature coefficients of inductance and capacitance cause designers few problems except when both components are connected together in *LC* oscillator circuits where frequency stability is important. One method of reducing frequency drift is to split the tuning capacitor into two sections, one of which has a negative coefficient. If a suitable combination of negative and positive coefficients is used the frequency drift can be reduced to well below 1 ppm.

Where the potentials in a circuit are determined by a resistor chain we have seen that the effect of temperature changes on these potentials will be very small. Matters are quite different, however, if the circuit includes semiconductor junctions. For silicon devices the current increases about 15% for a 1°C rise in temperature at constant voltage (Sparkes 1966). This means that a 20°C rise would cause an increase in current by a factor of just over 16. Since in an adverse environment the temperature may change much more than this, constant voltage operation of diode and transistor junctions is quite unacceptable. The usual method of stabilizing the transistor current against temperature changes is to connect a resistor in series with the emitter, chosen to ensure a voltage drop across it of least 2 V. The base voltage is held almost constant by a resistive potential divider. If the junction current is held constant, the base-emitter voltage falls by about 1.5 mV for each °C junction temperature rise. Thus for an 80°C rise, V_{be} falls by about 120 mV. For a fixed base voltage, the voltage across the emitter resistor will rise by the same amount, so causing an increase in emitter current by 0.12/2 or 6%. This is a rather crude calculation, but it is adequate to show the effectiveness of the method which is widely adopted. An alternative method of stabilizing emitter current is to make the base bias voltage fall with temperature at the required rate of about 2 mV per °C. This is often done by deriving the bias voltage from the voltage across a diode supplied with constant current. To obtain effective stabilization the diode and the transistor it regulates must be at the same temperature, generally arranged by mounting them on the same heat sink. This form of biasing is usually adopted for high-power amplifiers which have comparatively low-voltage supplies (typically 12 V for mobile operation). The 2 V dropped across an emitter resistor would then represent a significant power loss and a reduction in the effective voltage available for the transistor.

A final method of removing the effect of temperature changes is to isolate the circuit from them by enclosing it in an oven maintained at a constant temperature which must of course be above the maximum ambient temperature. The cost and power drain needed for this scheme means that it can be used in practice for only a small circuit package, typically a tuned circuit or a crystal used to determine the frequency of an oscillator. We would expect an improvement in frequency stability by an order of magnitude or more when using a constant temperature oven.

4.7 Design automation

Although some degree of automation is generally used in the design of electronic systems, it is largely confined to detailed activity such as the analysis of analogue and digital circuits and simulating their behaviour and assistance to the manufacturing process by helping the layout of printed circuit boards and integrated circuits. It is also used in the testing of the product at various stages.

Most of the programs used have been available for some years and nearly all of the faults in them have been discovered and removed. Despite this devices designed with their help still reveal occasional unexpected errors. The problem is that nearly all design aids involve at some stage computer programs which cannot at present be generated without some human effort, which in turn is likely to introduce errors. Thus all design activity should assume that errors will be present initially and some procedure for finding and correcting them is necessary.

The usual recommendation is to hold regular audits or reviews of the design, preferably conducted by engineers not involved directly in the design process. Experience shows that the designer is likely to overlook a mistake if he or she conducts the review him- or herself. This follows the advice given to authors that they should ask someone else to proof-read their work.

The problem of eliminating design errors has become of increasing interest as hardware has become more reliable and more faults are attributed to design and fewer to components. Although formal methods of designing systems are being developed (Diller 1990) they are as yet unable to tackle complex logical devices and are not in use commercially. The problem of exhaustive testing of intricate devices such as microprocessors lies in the large number of combinations of data, instructions and storage locations to be investigated. For example, even a small 8-bit microprocessor containing perhaps 70 000 transistors has many instructions and can address over 60 000 storage locations; a multiplier handling two 16-bit integers will have over 4 billion different input combinations. If every instruction is to be tested with all possible data values and all storage locations the test will take some hundreds of years. Thus only limited testing is practicable and it is important to design the tests to cover as much of the device logic as possible.

As microprocessors are now embodied in most military equipment, there is much interest in producing reliable devices without design errors. This can be largely overcome by using formal mathematical methods to specify and verify the processor. A team at RSRE Malvern has been working on this project, using a formalism called LCF–LSM (Logic of Computable Functions–Logic of Sequential Machines) and leading to a device called Viper (Verifiable Integrated Processor for Enhanced Reliability) intended for safety-critical applications (Dettmar 1986). This has a 32-bit data bus and a 20-bit address bus, and to avoid possible timing problems there is no provision for interrupts. All external requests for service are dealt with by polling; this can take longer than an interrupt if many devices are connected to the microprocessor, but with a fast processor and the moderate response time acceptable for servicing mechanical systems no problems arise. The only commercial use of the Viper device reported is in signalling equipment for the Australian railway network.

4.8 Built-in test equipment

Where equipment can be maintained availability can be increased by conducting regular system checks so that any fault is discovered as soon as possible. This enables repairs to be started as soon as possible, so minimizing the down-time. Two methods have been used, initial testing and periodic checking. Initial testing is usually included in single instruments such as high-bandwith oscilloscopes and logic-state analysers. These generally incorporate microprocessors to control their functions and are configured so that each time the equipment is switched on an interrupt is created which starts a test routine. This checks the calibration of the system and as many of its functions as possible. As this type of apparatus is generally used intermittently and hardly ever left running continuously, it is tested often enough to ensure that faults cannot give incorrect readings for very long.

Periodic testing is necessary for critical systems which are normally energized continuously and so would only have an initial test very occasionally. At regular intervals the system is diverted from its normal task and enters a test routine which conducts a quick system check, reporting any fault discovered. In large installations further diagnostic tests can then be carried out which will investigate the fault in more detail and give more information about its location.

Built-in tests of this kind are used in non-maintained systems only if some redundancy is provided and there are facilities for disconnecting faulty equipment and switching in alternative units.

4.9 Sneak circuits

A problem in some situations is the occurrence of what has been called 'sneak' circuits. These have been defined as latent paths or conditions in an electrical system which inhibit desired conditions or initiate unintended or unwanted actions (Arsenault and Roberts 1980). The conditions are not caused by component failures but have been inadvertently designed into the system. They are liable to occur at interfaces where different designers have worked on two packages but there has not been sufficient analysis of the combined system. They are also liable to occur after design modifications have been introduced when the new configuration has not been exhaustively tested.

One frequent source of sneak errors is the arrival of several signals required for particular action in an unexpected order. A similar source of potential error was recognized some years ago with the development of electronic logic circuits. If an input change causes more than one signal to propagate through a logic network, and two or more of these are inputs to the same gate, the resulting action can depend upon the order in which the various inputs arrive. If the output of the logic gate should not change it may nevertheless emit a short unwanted pulse which could advance a counter and cause an error.

This phenomenon is called a race hazard (Bannister and Whitehead 1983) and it must be avoided if the system is to operate reliably. It can be tackled in two main ways. The first generates what is called a 'masking' signal as an extra input to the gate which prevents a false output regardless of the timing of the input signals. This is satisfactory where the race can occur at only very few gates. In more complex systems such as digital computers it may occur many times and the effort of analysing these and introducing the extra logic is prohibitive. The solution adopted in this case is to inhibit the output of the gate until one can be certain that all inputs have arrived. The inhibiting signal is usually a train of constant frequency or clock pulses which is applied to all storage elements. Any inputs which arrive between clock pulses are not allowed to alter the state of the storage device until the next clock pulse arrives.

The procedures adopted to deal with race hazards in logic circuits can to some degree be applied to the prevention of the unwanted consequences of sneak circuits. Some of these may occur through the incorrect state of logic elements when power is applied to a package. Most logic devices which include some storage may set themselves in either logic conditions when power is first applied. In order to ensure that they all start operation from some known condition an initializing pulse is usually sent to them a short time after power is applied to set them into the desired state.

The same process can be used to avoid sneak circuits by delaying any action until one can be sure that all changes have occurred.

Extensive computer programs are now available to analyse systems and discover any sneak paths. They were originally written to handle electrical control circuits including relays, switches, lamps, etc., and were subsequently extended to apply to digital logic circuits. One event which helped to stimulate NASA to invest in sneak circuit analysis programs occurred at the launch of a Redstone booster in 1986. After 50 successful launches, a launch sequence was started but after lifting several inches off the pad the engine cut out. The Mercury capsule separated and ejected its parachutes, leaving a very explosive rocket on the pad with no means of control. It was left for just over 24 hours until the liquid oxygen had evaporated and the batteries had run down before being approached. Subsequent investigations showed a timing error had occurred in that the tail plug cable had disconnected 29 milliseconds before the control plug cable and the sneak circuit caused the engine to cut out. The cables were intended to disconnect in the reverse order and the cable arrangements were later altered to ensure this.

An unwanted digital input can occur if unused inputs to logic gates are not connected to either a logic 1 or logic 0 potential. Manufacturers always advise users to do this to prevent the inputs from picking up stray noise pulses. In one recorded case (Brozendale 1989) an unused input in an interface unit of a chemical plant was not earthed as intended and picked up an induced voltage. This caused an incorrect address to be sent to the controlling computer which gave the output commands for the wrong device. The result was that a number of valves were opened wrongly, breaking a gas line and releasing a toxic gas. Since plant safety depends critically upon correctly identifying the device which needs attention a safe system design should include more than one means of identification. The design principle is that all information exchanges between the processor and peripheral devices should have some degree of redundancy in the interests of reliable operation. Thus in addition to checking items such as addresses, it is desirable to read back into the computer all data sent to the peripheral devices so that it can be checked.

Bibliography

Arsenault, J. E. and Roberts, J. A. (1980) *Reliability and Maintainability of Electronic Systems*, Computer Science Press, New York

Bannister, B.R. and Whitehead, D.G. (1983) *Fundamentals of Modern Digital Systems*, Macmillan, London

Billinton, R. and Allan, R. N. (1983) *Reliability Evaluation of Engineering Systems, Concepts and Techniques*, Pitman, London

Brown, D. B. (1976) *System Analysis and Design for Safety Systems Engineering*, Prentice Hall, London

Brozendale, J. (1989) A framework for achieving safety-integrity in software. *IEE Conference Proceedings*, No. 314

Dittmar, R. (1986) The Viper microprocessor. *Electronics & Power*, October, p. 723

Diller, A. (1990) *An Introduction to Formal Methods*, Wiley, London

Henley, E. J. (1981) *Reliability Engineering and Risk Assessment*, Prentice Hall, Englewood Cliffs, NJ

Houpis, C. H. (1985) *Digital Control Systems: Theory, hardware, software*, McGraw-Hill, New York

Institution of Electrical Engineers (1979) Submarine telecommunication systems. *IEE Conference Proceedings*, No. 183

Jones, J. V. (1988) *Engineering Design, Reliability, Maintainability and Testability*, Tab Books, Blue Ridge Summit, PA

Kapur, K. C. (1977) *Reliability in Engineering Design*, Wiley, New York

Matthews, N. O. (ed.) (1974) *Introduction to Automated Testing*, Network, Newport Pagnell, Bucks

Osbourne, S. (1980) Developments in high reliability resistors. *Electronic Engineering*, March

Petroski, H. (1985) *To Engineer is Human: The role of failure in successful design*, Macmillan, London

Phillips, C. L. (1984) *Digital Control System Analysis and Design*, Prentice Hall, London

Sparkes, J. J. (1966) *Junction Transistors*, Pergamon, Oxford

Starke, C. V. (1988) *Basic Engineering Design*, Edward Arnold, London

Wobschall, D. (1987) *Circuit Design for Electronic Instrumentation: Analogue and digital devices from sensor to display*, 2nd edn, McGraw-Hill, London

5 Building high-reliability systems

5.1 Reliability budgets

Few electronic systems are designed for which no reliability target exists. This may vary from 'No worse than the opposition' for a mass-produced domestic article to a closely specified minimum MTBF for an avionic or military system, with perhaps a financial bonus for exceeding the minimum.

In the past some designs have been largely completed before an estimate of reliability was started. If this diverges significantly from the target a major redesign is required involving much extra time and cost. Consequently it is now accepted as a principle of good design that where reliability is a significant item in the specification, it should be a major consideration at all stages of the design.

In a system which can be regarded as a number of separate units, each of which must operate correctly if the system is to deliver its required output, it is useful to establish a reliability budget. In this the required overall reliability is partitioned between the various units so that the designer of each unit has his or her own reliability target.

A simple initial procedure which can be applied where the units have roughly the same complexity is an equal division. Thus if the overall reliability figure specified is R, the reliability target for each of n units is $\sqrt[n]{R}$. Thus for four units the target would be $\sqrt[4]{R}$. If the units vary in complexity the allocation should be unequal; a value can be assigned to each unit based upon previous experience or, given some preliminary design data, on a count of components, semiconductors or integrated circuit packages. The overall specification requires a relation between the unit reliabilities R_1, R_2, R_3, R_4 and R of

$$R = R_1 \times R_2 \times R_3 \times R_4 \tag{5.1}$$

It is more likely that the system will be specified as having a particular MTBF of M hours. In this case a system comprising four similar units will require each unit to have an MTBF of $4M$. Where the units have differing complexities and are expected to have differing MTBFs M_1, M_2, M_3 and M_4, the relation between them must be

$$1/M = 1/M_1 + 1/M_2 + 1/M_3 + 1/M_4 \tag{5.2}$$

An initial estimate of the various MTBFs can be made using a simple parts count and refined later as the design proceeds.

5.2 Component selection

Electronic components have been developed over many years to improve their performance, consistency and reliability and consequently the less demanding reliability targets can often be attained by using widely available commercial components. A modest improvement in reliability can in these cases be obtained by derating the components. There still remain, however, many applications for which high reliability is demanded and which thus need components of higher and well-established reliability.

Attempts to improve the reliability of electronic equipment were first tackled in a systematic way towards the end of the Second World War, when the American services discovered that some of their equipment cost during its lifetime at least 10 times more to maintain than to purchase initially. An early outcome of this was a programme to develop more reliable thermionic valves, which were responsible for many failures.

Work started in this country some 30 years ago when an increasing number of agencies such as the Post Office, commercial airlines, the armed services and the railways required very reliable electronic systems and attempted to issue their own specifications for component performance and reliability. The manufacturers found great difficulty in coping with all these different requirements and the long testing programmes needed for them. In consequence a committee chaired by Rear-Admiral G. F. Burghard was established to develop a set of common standards for electronic parts of assessed reliability suitable for both military and civilian applications. The committee's final report in 1965 was accepted by industry and government and the British Standards Institution (BSI) accepted responsibility for publishing the appropriate documents. The basic document is BS 9000 which prescribes a standard set of methods and procedures by which electronic components are specified and their conformance to specification is assessed. The system is implemented by the BSI and operated under the monitoring of the National Supervising Inspectorate.

BS 9001 gives tables and rules for sampling component production and BS 9002 gives details of all components which have been approved and their manufacturers. There is such variety in the items used to manufacture electronic equipment that separate specifications are needed for each family such as:

BS 901X Cathode-ray and camera tubes, valves, etc.
BS 907X Fixed capacitors
BS 9090 Variable capacitors
BS 9093 Variable preset capacitors
BS 911X Fixed resistors
BS 913X Variable resistors, etc.

In some cases, for example discrete semiconductors (BS 93XX), these have been divided into subfamilies such as signal diodes, switching diodes, voltage reference diodes, voltage regulator diodes, etc.

BS 9301 General-purpose silicon diodes
BS 9305 Voltage regulator diodes
BS 9320 Microwave mixer diodes (CW operation)
BS 9331 Medium-current rectifier diodes
BS 9364 Low-power switching transistors, etc.

There is now a European dimension to the BS 9000 scheme in that many British Standards are now harmonized with the standards of the European CECC (CENELEC Electronic Components Committee), and constitute the BS E9000 series.

Also, many CECC standards have been adopted as British Standards, for example:

BS CECC 00107 Quality assessment procedures
BS CECC 00108 Attestation of conformity
BS CECC 00109 Certified test records, etc.

The BS 9000 scheme includes provision for the collection of the results of life tests so as to build up a data bank of component performance.

5.3 The use of redundancy

Although greatly increased reliability can be obtained by using specially developed components, derating them and keeping them as cool and vibration free as possible there is a limit to the benefit this can bring. There are many critical applications where yet higher reliability is required and the usual method of coping with this requirement is to introduce some degree of redundancy.

In general terms this means providing more than one way of producing the desired output. The assumption is that if one path is inoperative due to a fault, another path will provide the correct output. This may involve sending copies of the same information along different paths (spatial redundancy) or sending copies of the information along the same path at different times (temporal redundancy). The former is much more powerful as it can cope with a permanent fault in one path, whereas the latter is generally simpler to implement but copes best with transient errors such as impulsive noise on a telephone line or a radio circuit. As only one path is provided it cannot cope with a permanent fault.

Many control and instrumentation systems are confined to a restricted area and so need no measures to cope with transmission faults. Thus the technique most applicable is that of spatial redundancy which requires extra equipment and if fully implemented the replication of the entire system. The simplest form of this is duplication. We postulate two identical channels, each of which can deliver the outputs needed, with some provision for switching to the spare channel when the working channel fails. If the probability of failure for a single channel is p, the probability of system failure is p^2 since the system will fail only when both channels fail. The reliability is thus

$$R = 1 - p^2 \tag{5.3}$$

This result assumes that both systems are independent so that the failure of one channel makes no difference to the probability of failure of the second channel. This assumption is not always valid, since any common item such as a power supply which feeds both channels will invalidate it. Even if two independent power supplies are provided, they will normally be connected to the same mains supply, and the result given in Equation (5.3) should be multiplied by the reliability of the mains supply.

We can generalize this result for n identical channels, on the assumption that only one working channel will provide the required output, to give

$$R = 1 - p^n \tag{5.4}$$

This again is not a realistic calculation since a multi-channel system will need some mechanism for checking the working channel and switching to the next channel when it fails. A better figure for overall reliability is given by multiplying R by the reliability of the checking and switching mechanism.

In some installations the checking may be done by a short program module in the computer which controls the system; this, however, may not be acceptable if it means suspending the computer's normal operation at regular intervals to run the test.

Where we have some reserve of data handling power and data appears in bursts we may be able to test a channel by injecting test signals between the bursts. This does not involve suspending the computer's operation, but is practicable only when we can be certain that there will be intervals in the demands for service which are long enough and frequent enough to allow adequate testing.

The program in such cases is usually divided into modules with varying degrees of importance which are executed, some at regular intervals of time, others when particular patterns of data occur. At any moment the program being executed is that which has the highest priority and when that has ended the module having the next highest priority is invoked. At any time the current program module can have its execution interrupted by another module of higher priority which becomes active. In the priority list the system test program is often put at the bottom so that when there is no other call for the computer's services it continues execution of the test program rather than idling.

5.4 Redundancy with majority voting

If we ignore the possibility of failure in the switching mechanism, and assume a constant failure rate for each channel of λ, the MTBF for a duplicate system is $M = 3/2\lambda$. For a triplicate system it increases to $11/6\lambda$ (Kapur and Lamberson 1977).

The scheme mentioned above means that the normal operation of the system must be halted for a short period to permit a channel to be tested. If it is found to be faulty it is disconnected and another channel switched into operation. This is bound to involve some interruption to the output, which cannot always be tolerated. For example, in a real-time vehicle control system any interruption can mean a major deviation from the desired path. In such circumstances some mechanism which does not involve a break in output is needed; one of these involves three identical channels fed with the same input signal with a majority voting circuit at the output. This is easiest to implement with a digital system where the required voting circuit has to implement the logical function $X = A.B + B.C + C.A$ where A, B and C are the outputs of the three channels. This requires only four gates as shown in Figure 5.1. The MTBF for this arrangement is $5/6\lambda$.

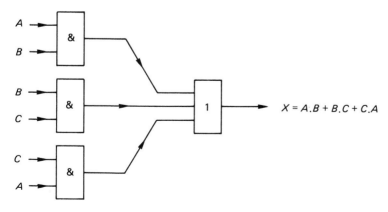

Figure 5.1 *Majority voting logic circuit*

It is important in maintained systems to provide an indication that a fault has occurred even though the redundancy has prevented this from causing a system failure. This enables corrective action to be started as soon as possible; without this the system is less reliable than a single system since a fault on either of the two remaining working channels will cause a system failure. The logic expression which must be implemented is derived easily by considering the outputs before a fault occurs. In this case the outputs must be either three ones or three zeros. The logic expression is thus $Y = A.B.C + \bar{A}.\bar{B}.\bar{C}$. A fault condition is indicated when this expression has a value zero. Thus to deliver a logic 1 signal when a fault has occurred we need the complement of this, that is

$$Z = \overline{A.B.C + \bar{A}.\bar{B}.\bar{C}} = B.\bar{C} + C.\bar{A} + A.\bar{B}$$

Although we have discussed majority voting in its simplest and most widely used form of triplicated channels, the same voting procedure can be used with any odd number of channels. The logic expression for the voting circuit is a little more complicated; for example, if we have five channels the terms comprise all of the combinations of three items selected from five, that is

$$X = A.B.C + A.B.D + A.B.E + A.C.D$$
$$+ A.C.E + A.D.E + B.C.D + B.C.E + B.D.E + C.D.E$$

This requires 10 three-input AND gates and one 10 way OR gate.

As an example of the benefit of triplication we take the control circuits of a recent optical fibre submarine cable. This has four separate channels, three working and one spare. At each repeater location the repeater inputs and outputs of each working channel can be switched to the spare channel in the event of a fault. If we assume that there are 200 components with an average failure rate of 0.2×10^{-9} per hour in the monitoring and switching operation at each repeater housing and we look for a working life of 20 years, the expected number of faults per housing is $200 \times 0.2 \times 10^{-9} \times 20 \times 8760 = 0.007\,008$.

The failure rate is a somewhat crude assessment as it is a weighted average over all the components, but the value is in line with the target failure rates quoted for a long-haul submarine repeater (Querol and Campagne 1980) which vary from 1.0 FIT for transistors to 0.1 FIT for capacitors and resistors (1 FIT is a failure rate of 10^{-9} per hour). The corresponding reliability is

$\exp(-0.007\,08) = 0.992\,945$

If there are 16 repeaters in the cable, the overall reliability of the switching operation will be

$0.992\,945^{16} = 0.8929$

If we introduce a triplicate redundancy scheme with majority voting at each repeater site the reliability will be

$$R_T = 3R^2 - 2R^3$$

where R is the reliability of each channel. This can be shown by considering the circumstances in which the system fails, that is when two channels are faulty and one working, or all three are faulty. If $p = (1 - R)$ is the probability of a channel failing, the probability of two or three failing is

$$P_T = 3p^2(1-p) + p^3$$
$$= 3p^2 - 2p^3$$

since $(1 - p)$ is the probability of one channel working and p^2 is the probability of two being faulty; there are three ways in which this can occur. Expressing this in terms of reliability gives

$$P_T = 3(1-R)^2 - 2(1-R)^3$$

Finally the overall reliability is given by

$$R_T = 1 - P_T$$
$$= 3R^2 - 2R^3 \tag{5.5}$$

Returning to the repeater calculation, the reliability of a triplicated version with majority voting is given by putting $R = 0.992\,945$ in Equation (5.5). The overall reliability then becomes $R_T = 0.9976$. Thus the probability of a failure has been reduced from 10.7% to 0.24%. This is a somewhat optimistic calculation since the reliability of the majority voting element has not been included. However, it should require far fewer components than the 200 we have assumed for each repeater station and thus should be much more reliable than the switching units.

5.5 The level of redundancy

The scheme shown in Figure 5.1 uses only one voting circuit, as the final element in the system. The overall reliability can be improved by subdividing the system, replicating each subsystem and following it by a majority voting circuit. It can be shown by using a somewhat simplified system model that the optimum scheme is one in which the system is subdivided so that the reliability of the subsystem is equal to the reliability of the voting circuit (Cluley 1981). Since in a digital system the same logic hardware is used in both the working channels and the voting circuit, the conclusion is that the subsystem and voting circuit should be of similar sizes. This is a practicable arrangement where discrete components are used, but most current equipment, both analogue and digital, makes much use of integrated circuits which generally have a much greater complexity

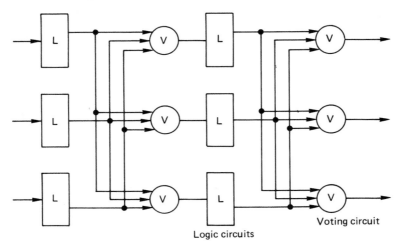

Figure 5.2 *Triplicated logic and voting circuits*

than a voting circuit. We are thus forced to conclude that optimum redundancy is impracticable in current equipment, and the number of voting circuits which can be introduced is limited by the system design. However, we can still obtain improved reliability by subdividing each channel and replicating the subassemblies; we can also ensure a further improvement by replicating the voting circuits so that each subassembly has its own voting circuit, as shown in Figure 5.2. The ultimate limit to the overall reliability is set by the final voting circuit, which cannot be replicated, although it could have some degree of component redundancy.

5.6 Analogue redundancy

The same increase in system reliability due to triplication which is obtained in digital systems can also be obtained in analogue systems. A practical difficulty is the design of suitable majority voting elements. One circuit which was developed for a triplicated analogue autopilot system for aircraft will give a majority vote if any two channels have the same output. If the outputs are all different, it will follow whichever output has a value intermediate between the other two. Thus it will give the desired output if one output is hardover to zero and a second is hardover to full scale; we assume that the third output is correct. If, however, two outputs both give either zero or full scale the circuit gives an incorrect output (Cluley 1981). Another arrangement was used in a later aircraft control system in which the three channels drove servo motors which rotated the shaft on which the control surface was mounted. The shaft summed the three torques generated by the motors and so achieved an approximate majority vote. To avoid damage to the motors or their driving amplifiers the motor current and hence the torque was strictly limited. The effective voting element was the control shaft which could easily be made large enough to ensure that the probability of its failure was negligible.

An alternative arrangement which is convenient to use with integrated circuit amplifiers is to operate them in parallel pairs, with provision for disconnecting a faulty amplifier. As these devices are directly coupled any fault will almost certainly disturb the potential at the output. This is fairly simple to arrange if the amplifier output is restricted. For example, if it does not exceed $5\,V \pm 2\,V$ with a $10\,V$ supply a two-diode circuit as shown in Figure 5.3 will disconnect the amplifier from the output when its potential falls below $3\,V$ or above $7\,V$. In the first case D1 disconnects and in the

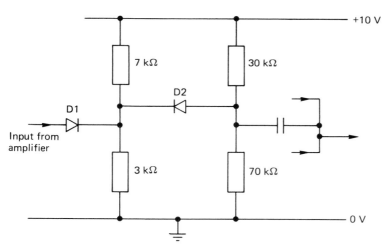

Figure 5.3 *Circuit to disconnect redundant amplifier when its output is greater than $7\,V$ or less than $3\,V$*

second case D2 disconnects. In practice the disconnection is not abrupt owing to the diode characteristic, but when the amplifier output is hardover to earth or the positive supply there will be nearly $3\,V$ reverse bias on one of the diodes, which is ample to ensure complete disconnection of the faulty amplifier.

Where a data signal can vary widely in amplitude in a random fashion much advantage can be obtained by combining several versions of the signal provided that the amplitude fluctuations are largely independent. This is the technique used in radio receivers for long-distance circuits. The received signals arrive after one or more reflections from the ionosphere and are liable to fluctuate in amplitude because of interference between waves which have travelled along different paths. Experiment shows that the fluctuations in signal level received by aerials spaced 10 wavelengths apart have very little correlation, and if they are combined on the basis that the largest signal is always used, the result will show much less fluctuation than any component signal. This technique is called diversity reception and is often used to combat fading in long-distance radio reception, particularly with amplitude-modulated transmissions. The amplitude of the received carrier can be used to indicate signal strength, and is used as a feedback signal to control receiver gain. In triple diversity the three gain-control signals are taken to three diodes with a common output which automatically selects the largest signal. This is connected to all three receivers and the audio outputs are also commoned. The gain of the receivers handling the weaker signals will be reduced and so their contribution to the common audio output will also be reduced. Although this technique was first used for HF reception it was also found to improve VHF reception from moving vehicles in telemetry links for instrumentation.

5.7 Common mode faults

A crucial factor in designing redundant systems is ensuring that all of the replicated channels are independent, so that the existence of a fault in one channel makes no difference to the probability of a fault occurring in another channel. Any fault which will affect all channels is called a 'common mode' fault and we can only obtain the full improvement in reliability which redundancy promises if common mode faults are extremely unlikely. Two likely causes of common mode faults are common power supplies and common environmental factors.

Where all the channels of a redundant system are driven from a common power supply, the reliability of this supply will be a limiting factor in the overall system reliability. Should this be inadequate, the power unit can be replicated in the same way as the data channel and its reliability can be included in the estimate of channel reliability. There is still a common mode hazard since all the power units are connected to the same source of energy. Past records give a useful indication of the probable reliability of the supply; if this is considered to be inadequate there are several ways of coping. The first is the use of uninterrupted power supply (UPS). This consists of a motor–generator set supplying the load which has a large flywheel mounted on the common shaft. When the main power supply fails a standby diesel engine is started and as soon as it has attained full speed it is connected via a clutch to the motor–generator shaft and takes up the load. While the diesel engine is starting the flywheel supplies the energy needed by the generator. The shaft speed will fall somewhat but this can usually be tolerated. Where long supply interruptions may occur a second diesel generator can be provided.

There are some variants of this: for low-power applications the supply may be obtained from a battery during a mains failure, the battery otherwise being trickle charged from the mains. For higher-power loads the main supply can be from a mains-driven rectifier, with a standby generator in the event of a mains failure. To allow time for the generator to run up to speed a standby battery is normally provided, sufficient to supply the load for 10-20 minutes. The scheme depends upon the switching generally performed by diodes which automatically connect the load to the highest voltage supply available. Thus the normal working supply voltage and the generator voltage must both be a little greater than that of the standby battery. The most likely failure mode of the power diodes used for switching is to a short circuit, so the reliability can be improved using twin diodes in series as shown in Figure 5.4.

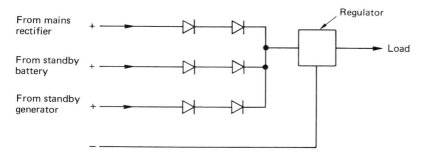

Figure 5.4 Redundant power supply switching

Other common factors which need addressing are the environment which is likely to affect all channels of a replicated system, and secondary damage. Since all adverse environmental conditions such as large temperature fluctuations, excessive vibration or a corrosive atmosphere reduce the reliability of electronic equipment their effect will be particularly severe if they affect all the channels of a redundant system. The effect of the environment may be diminished by enclosing the complete system in an insulating housing, but this will not completely remove the chance of common mode faults unless there is some segregation between the channels. For example, a fault in one channel may cause it to overheat or emit toxic fumes; without segregation between the channels this may cause a fault in the other channels.

Common mode faults of this kind have occurred in power stations where despite some degree of redundancy all of the control cables were routed along the same duct as some power cables. A cable fault once caused a fire which damaged all of the control cables, so nullifying the benefit of redundancy. The only way to avoid such common mode faults is to ensure complete physical separation not only between power and control cables, but also between the cables incorporated in each of the redundant channels.

With the increasing reliability of electronic hardware, a greater proportion of faults caused by design errors appear in maintenance records. These are likely to occur as common mode failures since the same error will arise in all channels of a replicated system. Although such systems are normally subjected to exhaustive testing, it is very difficult to ensure that a fault which occurs only with a particular set of data has been eliminated. The complexity of current computer-based systems means that they would be obsolete and out of production long before all possible combinations of data, instructions and storage locations had been tried.

One expensive but apparently worthwhile way of reducing the consequence of design faults in redundant systems is to use different teams to design the nominally identical channels. The extra expense has been thought beneficial in high-reliability applications such as satellite control systems, both for hardware and software design.

Bibliography

Arsenault, J. E. and Roberts, J. A. (1980) *Reliability and Maintainability of Electronic Systems*, Computer Science Press, New York

Ascher, H. and Feingold, H. (1984) *Repairable Systems Reliability*, Marcel Dekker, New York

Billington, R. and Allan, R. (1983) *Reliability Evaluation of Engineering Systems, Concepts and Techniques*, Pitman, London

Cluley, J. C. (1981) *Electronic Equipment Reliability*, Macmillan, London

Dhillon, B. S. (1983) *Systems Reliability Maintainability and Management*, Petrocelli, Princeton, NJ

Dittmar, R. (1986) The Viper microprocessor. *Electronics & Power*, October, p. 723

Green, A. E. (ed.) (1982) *High Risk Safety Technology*, Wiley, Chichester

Haifley, T. and Bhatt, A. (1987) Fault-tolerant ICs: The reliability of TMR yield-enhanced ICs. *IEEE Transaction on Reliability*, **R-36**, No. 2

Kapur, K. C. and Lamberson, L. R. (1977) *Reliability in Engineering Design*, Wiley, London

Kershaw, J. (1989) Dependable systems using 'VIPER'. *IEE Conference Proceedings* No. 314, p. 23

Querol, A. and Campagne, J. P. (1980) Philosophy of component selection for submerged repeaters. *IEE Conference Proceedings* No. 183

Wobschall, D. (1987) *Circuit Design for Electronic Instrumentation*, 2nd edn., McGraw-Hill, London

Yaacob, M. Y. et al. (1983) Operational fault-tolerant microcomputer. *Proceedings of the IEE*, **130**, Pt E, No. 3, May

6 The human operator in control and instrumentation

6.1 The scope for automation

Although for economic reasons control and instrumentation systems are increasingly being automated there remain many situations in which decisions have to be made with complex and sometimes incomplete data. In many of these situations a human operator is still included in the loop on account of his or her flexibility and ability to accept a wide range of data. Thus although it has been technically possible for many years to apply the brakes automatically on UK mainline trains when they pass a danger signal, railway managers have hitherto insisted on leaving this critical decision to the driver.

In some situations the control information is provided by a number of different sources which are not directly detectable by a human controller. For example, in the blind landing system used for civil aircraft the normal visual data the pilot uses is absent due to adverse weather, and information about height, course and position relative to the runway is provided by radio beams and the magnetic fields of cables in the ground.

Although it would be possible to convey this information to the pilot in a visual form the quantity and rate of change of the data would impose a very severe load and international regulations do not permit manual landing in these circumstances. The last few minutes of the landing are then controlled automatically.

In other situations where automation is technically possible a human operator is retained in the interest of safety on account of greater flexibility and the ability to cope with quite unexpected conditions. The postal service has run driverless underground trains across London for some decades, but they carry only freight. As yet there are very few miles of track which carry driverless trains for public use.

Most commercial passenger ships have automatic steering gear, but international safety legislation requires a suitably qualified person to be in charge of the ship whenever it is under way.

6.2 Features of the human operator

One way of analysing the dynamic behaviour of a control system is to determine its overall transfer function, by combining the transfer functions of all its component parts. This type of analysis thus requires a knowledge of the transfer function of a human operator, when he or she is an essential part of the control loop, for example as a car driver or an aircraft pilot. Unfortunately it is very difficult to define a satisfactory model of human behaviour as there are so many possible transfer functions which may be exhibited. Most operator errors are noticed and corrected by the operator, who is often unwilling to admit mistakes. Thus any attempt to measure performance must be sufficiently sensitive to detect these corrected errors and must at the same time be inconspicuous so as not to distract the operator. A further area of uncertainty is that performance depends upon the physical and mental state of the operator; it is very difficult to determine this and to specify it in any meaningful way.

Some attempts to estimate the reliability of a human operator have been made by breaking down his or her task into a series of simple operations and assigning a probability of error to each of these. Some estimates of error rates given by Hunns and Daniels (1980) are shown in the table below.

Type of error	*Rate*
Process involving creative thinking, unfamiliar operations where time is short, high-stress situation	10^{-0}–10^{-1}
Errors of omission where dependence is placed on situation cues and memory	10^{-2}
Errors of commission, e.g. operating the wrong button, reading the wrong dial, etc.	10^{-3}
Errors in regularly performed, commonplace task	10^{-4}
Extraordinary errors – difficult to conceive how they could occur; stress-free situation, powerful cues helping success	$<10^{-5}$

As some confirmation of these figures it is generally accepted that the average person will make an error when dialling a telephone number once in about 20 attempts. The error rate is, however, somewhat lower if push-buttons are used instead of a circular dial.

One factor which seriously affects the effectiveness of the human operator is the duration of the control task. In 1943 the RAF asked for tests to determine the optimum length of a watch for radar operators on anti-submarine patrol, as it was believed that some targets were being missed. Tests showed that a marked deterioration in performance occurred after about 30 minutes, and this conclusion has often been confirmed since.

It is important not to regard this particular result as directly applicable to other circumstances. A specific feature of the radar operator's task on anti-submarine patrol is the low data rate and the use of only one of the senses. Many control tasks involve much higher data rates and data input to several senses; for example, an activity which many people find comparatively undemanding – driving a car – involves visual and audible input and sense feedback from the pedals and the steering wheel.

There is some evidence that a human operator can handle a greater data rate if several senses are involved, rather than using a single sense. For reliable control action it is important to choose the appropriate sense to convey data to the operator. Visual communication is best where the message is long, complex or contains technical information, where it may need referring to later, or where it may contain spatial data.

On the other hand audible communication is more effective for very simple, short messages, particularly warnings and alarms, or where the operator may move around. In a crowded environment the operator may easily move to a position where a display cannot be seen, whereas an audible warning would be heard. Also, audible information is preferable where precise timing is required to synchronize various actions; for example, an audible countdown is always used when spacecraft are launched although ample visual displays are always available.

Despite its many advantages speech communication is liable to errors; the average human operator has a vocabulary of some tens of thousands of words and it is easy to confuse one with another. A method often used to make the communication more reliable is to diminish the number of words which can be used by confining the message as far as possible to a set of standard phrases. This is a procedure used successfully by the services and air traffic control. Ideally each message should be read back to enable the originator to check its contents, but the rate at which information must be handled often precludes this. Sometimes there are alternative sources of information; for example, in air traffic control the radar display enables the controller to check that the pilot has understood his or her message and is flying on the requested bearing. Mistakes do still occur, however, and these may have serious consequences. In 1977 two Boeing 747 aircraft collided on the ground at Tenerife with the loss of 583 lives and a cost of some $150 million. A number of human errors contributed to this accident, a major one being a misunderstanding between the air traffic controller and the pilot of one aircraft. The pilot thought that he had been cleared to take off, whereas he had been cleared only to taxi to the end of the runway and was expected to request further clearance when he reached there.

Tests on the efficiency with which simple tasks are performed show that this depends upon the rate at which data is presented to the operator. If the data rate is too slow attention slackens and performance suffers. Thus the designer of a control system which involves a human operator should ensure that the rate at which information is presented to the operator must be high enough to keep him or her alert, but not too high to overload his or her capacity to accept and comprehend it. The period for which a control task can be continued before performance begins to deteriorate is of course dependent to some degree upon the task and the individual, but experiments suggest that regular breaks will improve the reliability of the operator.

One outstanding feature of the human operator is his or her adaptability; an operator is able to alter his or her working pattern fairly quickly when presented with a new machine or working environment, and can cope with quite complex data. For example, an operator can read handwritten information comparatively easily despite its variety of size and character; computers cannot yet perform this task reliably. An operator can also learn to handle new machines and procedures comparatively quickly. However, if the old procedure has been in place for some time and has become very familiar, there are many cases in which, under stress, the operator will revert to his or her earlier behaviour. Records of aircraft accidents and near misses contain many examples of this behaviour (Chapanis 1965).

When a human operator forms part of a control process it is important to have some data about the operator's response time. Measurements of this have been made when various tasks are being performed. A relation quoted (Schooman 1968) between response time t and the display information H in bits is

$$t = a + bH \text{ seconds} \tag{6.1}$$

Here a is the lower limit of human response time, equal to 0.2 second, b is the reciprocal of the information handling rate, typically 15 bits per second. The task performed was monitoring a set of instruments and operating toggle switches when certain readings occurred.

H is derived from the display by calculating

$$H = \log_2 n$$

where n is the number of equiprobable, independent, alternative readings of the display.

Although one particular advantage of including a human operator in a control system is adaptability, this may in some cases cause problems. In any complex system the operator needs a degree of autonomy to enable him or her to cope with unexpected situations, but it is important that this is not used to cut corners and thus to allow dangerous situations to arise. It is a natural human characteristic that although an operator may start a new task carrying out all the instructions correctly, without careful supervision the operator tends to relax and discover ways of easing the task and perhaps defeating some of the in-built safety provisions.

6.3 User friendly design

In many early control panels much of the designer's work was directed towards producing a balanced and often symmetrical layout of the instruments and controls. Too often the needs of the operator were ignored; the operator had to stretch out to reach some switches and meters were not always in a convenient position to read quickly and accurately. The result was that in times of stress or emergency operators made errors, and in order to minimize these a systematic study was made of the human frame to determine where switches and dials should be placed to be within convenient reach, the position and intensity of lighting, and other similar matters which would ease the operator's task. Many of the arrangements proposed as a result of the study were contrary to previous ideas for a balanced and symmetrical layout.

For example, to help pilots to identify the particular control knob they needed to operate when these were grouped closely together, each knob was given a different shape and surface texture. Also, in control panels used for chemical plant and power stations a simplified line diagram of the plant was sometimes printed on the panel, with the controls and instruments connected to each unit of the plant near to the symbol for that unit.

Many control activities require the operator to turn a knob or move a lever to bring the reading on a meter to a target value. In this situation it is important to ensure that the meter and knob are mounted close to one another and that there is proper coordination between the direction in which the knob is turned and the direction in which the meter pointer moves. Normally a movement of the pointer from left to right should be caused by a clockwise rotation of the control knob. This is satisfactory with the knob beneath the meter or to its left or right. Another factor needing attention in this situation is the amount of pointer movement caused by a given rotation of the knob. If this is too small control action is delayed because the knob may need several turns; on the other hand this arrangement assists the exact setting of the control. Several schemes are used to minimize the delay: two gear ratios may be provided – fast for rapid movement and slow for fine adjustment – or a flywheel may be mounted on the control shaft so that a rapid spin will continue to cause rotation for some seconds.

On the other hand if a small knob rotation causes a large deflection of the pointer it will be difficult to set its position accurately and this may take longer than necessary.

In situations where rapid, almost instinctive action may be needed in an emergency, for example driving a car or piloting an aircraft, it is important to have the most frequently used controls in standard positions. For some 70 years the positions of brake, clutch and throttle have been standardized on most of the world's manually operated cars, so preventing many potential accidents which could arise when drivers use unfamiliar cars. Unfortunately the same is not true for cars with automatic gearboxes and a number of accidents have been caused by variations in the placing of the controls and their labelling.

Such variations were much more liable to cause errors in flying aircraft, and some research was conducted soon after the Second World War into events which had been classed as 'pilot errors'. This research revealed that in many of these events the awkward layout of the controls and the confusing displays were a contributing factor to the accident. A particular example of this (Chapanis 1965) is the position of three important controls which are mounted on the throttle quadrant in three American military aircraft then in common use. These are as shown in the table below.

	Position on throttle quadrant		
Aircraft	*Left*	*Centre*	*Right*
B-25	Throttle	Propeller	Mixture
C-47	Propeller	Throttle	Mixture
C-82	Mixture	Throttle	Propeller

Clearly pilots who move from one aircraft to another are very likely to confuse these controls in an emergency and several cases of this were uncovered.

An unusual case of failure to interpret instrument readings correctly occurred in the crash of a Boeing 737-400 at Kegworth in January 1989, which killed 47 people. The official accident report said that a fan blade fracture had caused a fire in the left engine, causing smoke and fumes to reach the flight deck. Together with heavy engine vibration, noise and shuddering the resulting situation was outside the pilot's experience as they had not received any flight training for the recognition of engine failure on the electronic engine instrument system. The captain later told investigators that he had not obtained any clear indication of the source of the problem when he looked at the instrument panel, and the co-pilot had no recollection of what he had seen. However, evidence from the flight data recorder and the remains of the aircraft show clearly that the crew shut down the right engine instead of the left engine so depriving the aircraft of all power. They mistakenly assumed that they had behaved correctly since after shutting down the right engine the noise and shuddering from the left engine ceased. The complete loss of power meant that the aircraft could not reach the runway at which the captain hoped to land, and it crashed on the edge of the M1 motorway. Although many people on board, including the cabin crew, had seen flames coming from the left engine, this information did not reach the pilots.

It is difficult to understand how two experienced pilots could both misinterpret the engine instruments so fatally as to shut down the wrong engine, but subsequent analysis of the engine instrument panel revealed major deficiencies in its layout. The primary instruments for both engines were in a panel on the left and the secondary instruments in a panel on the right. Each panel was aligned with a throttle control as shown in Figure 6.1 (Goold et al. 1990).

The clearest warning of the engine failure was given by the vibration indicator, which is a secondary instrument and thus in the right group and nearest to the right throttle. In the confusion the right engine was throttled down. The accident report suggested an alternative layout also shown in Figure 6.1. Here the primary instruments are aligned with the throttle controls and the primary instruments are aligned with the throttle controls and the secondary instruments are on either side of them. The Royal Air Force Institute of Aviation Medicine (IAM) carried out tests after the crash on the 737-400 instrument layout to find out how it was possible for the pilots to misinterpret the information presented to them. The IAM found that the actual combination of layout and dial design was the least efficient of the four combinations tested and gave 60% more reading errors than the best and took 25% longer to read. Also pointers pivoted at their centres were found to be much more effective than the cursors fitted (Learmont and Moxon 1991). These were small light-emitting diode displays which moved round the edge of a circular scale.

The main display showed fan speed, exhaust gas temperature, core speed and fuel flow. In addition to the analogue display given by three LEDs a digital display was given in the centre of the device. The secondary displays were a little smaller and showed oil pressure, oil temperature, and A and B system hydraulic pressures. These were analogue only. The main display LEDs have 81 different positions and the secondary displays have 31.

A useful indication of normal conditions is the same reading for each quantity of the two engines. With the earlier pointer display this was clearly shown by the parallelism of the two pointers. It is considerably more difficult to detect this from the position of three small LEDs outside the marked scale of the electronic display.

In order to assist pilots to recognize warning indications correctly, all modern cockpits now adopt the dark/quiet principle, meaning that if all

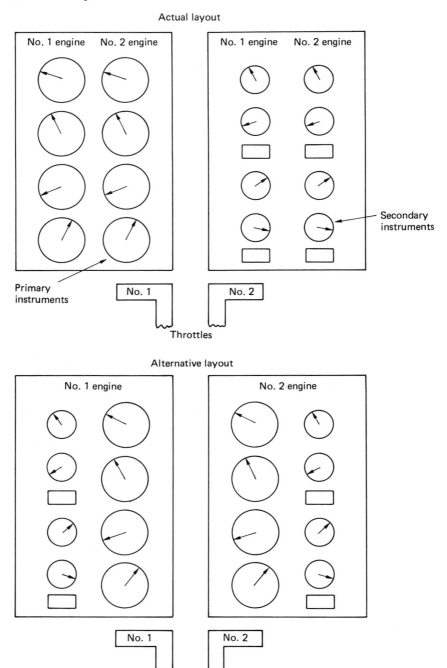

Figure 6.1 Layout of engine instruments in Boeing 737-400 aircraft

systems are working as they should be there will be no lights or sound. In consequence any warning given by light or sound is instantly recognized. This arrangement is used in all the Airbus models and is preferred by pilots.

The instrument panel in the Airbus 320 differs considerably from that of earlier aircraft in that much of the data is presented on cathode-ray tube displays. This allows much more information to be available to the pilot since he or she can switch the CRT to several different sources of data. This facility is a valuable help in fault finding. There is, however, one point in which the cockpit arrangements differs from those of previous aircraft: when the engines are under automatic control the throttle levers do not move when the engine power is altered, although they are operated as normal when under manual control. In other aircraft the throttle controls move when under automatic control so giving the pilot an indication of the power demanded. Pilots have become accustomed to having this information and find that it takes longer to find by looking at the instrument display.

6.4 Visual displays

An important factor in the design of warning systems is the amount of information which is presented to the operator. This is important both in total and in each separate display. It is usually recommended that in a text display not more than 25% of the available space should be used. If the text is packed more tightly it becomes more difficult to read and understand. The same general principle applies to diagrams; if these are too crowded they become much more difficult to interpret.

The eye is most sensitive at illumination levels typical of electronic displays to light of about 550 nanometres wavelength, a yellow/green colour, so this is advised for monochrome displays. This view is supported by Ellis *et al* (1977), who report tests in which the time the subject took when asked to read the first character in a four-digit display was measured using light of various colours. Other tests on the use of passive displays which reflect or scatter incident light and active displays which emit light, such as LEDs, were conducted. These indicated that both types were equally readable in high levels of illumination, but the active display was much easier to read in low light levels such as exist in the less well-lit parts of the cockpit.

There is a limit to the rate at which humans can read and assimilate information and if this is presented by too many warnings and indicators the result is confusion and often incorrect responses. In some complex control environments the designers attempt to provide a warning for almost all abnormal conditions, and in an emergency the operator may be overwhelmed with visual and audible signals. Current civil airliners are in danger of suffering from this excess; for example, there are 455 warnings and caution alerts on the Boeing 747 (Hurst and Hurst 1982).

A notable example of operators being overwhelmed by an excess of alarms occurred in the American nuclear power station on Three Mile Island in March 1979. At 4 a.m. a pump feeding water to the steam generators failed. The emergency pumps started but could not function properly because outlet valves which should be open were closed. As a result water backed up into a secondary loop causing a rise of pressure in the reactor. A relief valve then opened automatically, but stuck open when it should have reclosed as the pressure fell. The emergency core cooling system was activated and water poured into the reactor and out through the relief valve. A further fault in the instrumentation caused the operators to assume that the reactor was covered with water and so safe, whereas the water was running out. The operators then took over and shut down the emergency cooling system, so depriving the reactor of all cooling. The fuel rods then soon reached a temperature of 2500°F. This complex set of events produced so many alarms and flashing lights that the operators were completely confused and it was nearly two hours before they realized the real problem (Moss and Sills 1981, Rubinstein 1979).

6.5 Safety procedures

Although the benefits of including a human operator in any complex control system are well recognized – the greater flexibility of action and the ability to tackle situations not previously encountered, for example – there may well be accompanying hazards. One feature of humans is the change in attitude caused by carrying out a fairly well-defined control task many times. Initially a new task tends to be undertaken carefully with due attention to all the prescribed checks and data recording. With time, however, there is a temptation to cut corners and look for the easiest way to achieve a marginally acceptable result, and unless strict supervision and monitoring are insisted upon some safety procedures may be neglected. In the design of high-voltage switching stations provision must be made to isolate some parts of the system to permit routine maintenance and extensions, and it is essential that the power should not be reconnected until work has finished and the staff have left the scene. Although there are administrative procedures to ensure this, such as 'permit to work' cards, there is also a physical interlock system. This requires the power to be switched off before a key is released which gives access to the working area. When the gate to the working area is open, the key is trapped and cannot be used to turn on the power.

The high-voltage areas of radio transmitters and nuclear installations are normally fenced off and the interlock is usually a switch fitted to the access doors. When these are opened the power is automatically disconnected and cannot be reconnected without moving outside the area. Unfortunately there are occasionally undisciplined operators who try to make adjustments with the power on by climbing the protective fence with fatal consequences.

A particular accident in which human errors were a major factor was the explosion in the Chernobyl nuclear reactor in 1986. Several of the people in senior positions in the power station had little experience in running large nuclear plant, and the prevailing attitude of secrecy prevented the details and lessons of earlier accidents reaching plant operators and managers. Even the information about the Three Mile Island accident was withheld, although the rest of the world knew all about it. In addition those in charge of the industry put much more emphasis on production than on safety. The experiment which was being conducted when the accident occurred was to test the ability of the voltage regulator to maintain the busbar voltage when the turbines were slowing down after the reactor was shut down in an emergency. The voltage needed to be held near to normal for 45-50 seconds when supplying the essential load to ensure that safety mechanisms were effective, including the emergency core cooling system. The regulators on other similar stations had given some trouble and many proposals had been made to conduct the same kind of test. However, this was considered risky and the plant managers had refused to allow it. The management at Chernobyl had nevertheless agreed to the test; it is unlikely that they fully realized the possible consequences of their decision.

Bibliography

Apostolakis, G. E. and Bansal, P. P. (1977) Effect of human error on availability of periodically inspected redundant systems. *IEEE Transactions on Reliability*, **R-26**, 220
Beatty, D. (1991) *The Naked Pilot: The human factor in aircraft accidents*, Methuen, London
Booth, P. A. (1989) *An Introduction to Human-Computer Interaction*, Erlbaum, London
British Nuclear Energy Society (1987) Chernobyl, a technical appraisal. *Seminar Proc. Publ. London for the BNES*
Brown, C. M. (1988) *Human-Computer Interface Design Guidelines*, Ablex, Norwood, NJ
Burgess, J. H. (1989) *Human Factors in Industrial Design*, TAB Books, Blue Ridge Summit, PA
Chapanis, A. (1965) *Man-Machine Engineering*, Tavistock Publications, London
Ellis, B., Wharf, J. H. and Tyte, R. N. (1977) Modern displays for cockpit applications. *IEE Conference Publication* **150**, 35
Forman, P. (1990) *Flying into Danger: The hidden factors about air safety*, Heinemann, London
Gayler, (1987) *Applied Ergonomics Handbook* 2nd edn, Butterworth, London
Goold, I. *et al*, (1990) *Flight International*, 31 October, p. 24
Hawkins, F. H. (1987) *Human Factors in Flight*, Gower Technical, Aldershot
Henneman, R. L. and Rouse, W. B. (1984) Human performance in monitoring and controlling hierarchical large-scale systems. *IEEE Transactions on Systems, Man and Cybernetics*, **SMC-14**, 184

Hensley, G. (1983) Human reliability assessment. *The Chemical Engineer*, July, p. 9
HMSO Air Accident Investigation Board (1990) Report on Accident to Boeing 747-400 G-OBME near Kegworth, Leics on 8 Jan 1989. *Air Accident Report* 4/90
Hunns, D. M. and Daniels, B. K. (1980) *The Method of Paired Comparisons*, 6th Advances in Reliability Technology Symposium, UK AEA, Warrington
Hurst, R. and Hurst, L. R. (eds) (1982) *Pilot Error: The human factors*, 2nd edn, Granada, London
Kletz, T. A. (1985) *An Engineer's View of Human Error*. Institute of Chemical Engineering, Rugby
Learmont, D. and Moxon, J. (1991) *Flight International*, 27 November, p. 25
Lee, K. W. *et al.* (1988) A literature survey of the human reliability component in a man–machine system. *IEEE Transactions on Reliability*, **R-37**, 24
McCormick, E. J. (1987) *Human Factors in Engineering and Design*, 6th edn, McGraw-Hill, London
Martinez, J. M. *et al.* (1990) An analysis of the physical causes of the Chernobyl accident. *Nuclear Technology*, **90**, 371
Butterworth, London
Moss, T. H. and Sills, D. L. (eds) (1981) The Three Mile Island nuclear accident, lessons and implications. *Annals of the New York Academy of Sciences*, **365**
Rubinstein, E. (1979) Special Issue: Three Mile Island and the future of nuclear power. *IEEE Spectrum*, **16**, 30
Schooman, M. L. (1968) *Probabilistic Reliability: An Engineering Approach*, McGraw-Hill, New York
Schum, D. A, and Pfeiffer, P. E. (1973) Observer reliability and human inference. *IEEE Transactions on Reliability*, **R-22**, 170

7 Safety monitoring

7.1 Types of failure

Hitherto we have considered that a fault in a component occurs when its characteristics change enough to prevent the system in which it is incorporated producing the expected output. The only categories which have been recognized are those dealing with changes with time – permanent or intermittent faults – and those dealing with the rate of change – degradation or catastrophic failures. For example, both open-circuit and short-circuit failures are regarded equally as catastrophic failures. This classification is valid for all data handling systems, but needs further refinement for some control systems and safety monitoring systems.

In these we can envisage two consequences of failure, depending upon the state in which the system is left afterwards. In safety systems which are used to monitor some process, plant or machinery the important process parameters are measured continuously or at frequent intervals. When any parameter departs from its target value so far as to constitute a safety hazard, the safety system should shut the process down or ensure that it moves to a safe condition.

The safety system then has only two possible output states: it can shut the system down, as it is required to do if a fault exists, or it can do nothing and allow the process to continue, as it should do if there is no fault. A fault in the monitoring equipment can then have two consequences: what is called a 'fail-safe' error occurs when the system is shut down although no fault exists, and a 'fail-dangerous' error occurs when the monitoring function fails so that the system will not be shut down when a plant fault occurs.

Generally the consequence of a fail-safe error is much less than that of a fail-dangerous error. A fail-safe error normally interrupts plant operation, and so will reveal itself immediately when the plant is operating. If the plant is shut down, the fault will be detected only when an attempt is made to bring the plant into operation. The shut-down process is usually designed so that no damage will be caused to the plant or its operators.

A fail-dangerous error may cause serious damage to the plant and operators. Without routine maintenance a fail-dangerous error on the safety monitoring equipment will manifest itself only when a plant fault occurs; the expected automatic shut-down will then not occur.

The most severe consequence of a fail-dangerous error lies probably in the safety monitoring circuits of a nuclear power reactor. A fail-safe error will cause an unscheduled shut-down of the station, costing possibly some hundreds of thousands of pounds per hour of down-time. A fail-dangerous error may at worst cause a calamity on the scale of the Chernobyl disaster with a cost which is very difficult to quantify, but which will certainly be very large. An American investigation in 1957 led to the Brookhaven Report (WASH-740) which estimated that the property loss in a worst-case release of radioactivity could be $7 billion. A later survey in 1964-5 took into account price rises and inflation and revised the total cost to some $40.5 billion (Curtis and Hogan 1980).

Generally the design philosophy will be to equalize approximately the damage potentials of safe and dangerous failures. The damage potential is defined as the product of the probability of the failure and its effect (Klassen and van Peppen 1989). The most convenient measure of the effect is its cost. This is often difficult to estimate, particularly if loss of life is involved, but is necessary to establish a justifiable design. Some international agreements may be helpful; for example, there is an agreed maximum payment for loss of life in accidents involving civil airliners.

The cost of a major disaster at a nuclear power station will certainly be much greater than the cost of a safe failure, so the probability of a dangerous failure must be reduced far below that of the probability of a safe failure. This usually involves a high degree of redundancy in the safety system. A proposed target figure for the average probability of failing to meet a demand to trip is 10^{-7} over 5000 hours (Jervic 1984).

7.2 Designing fail-safe systems

Any attempt at designing an electronic monitoring system which is much more likely to fail safely than fail dangerously immediately raises the question of fault data. Ideally we need to discover the most likely failure mode of all components and design our circuits so that this type of failure leaves the system being monitored in a safe condition. For many components the two basic types of failure are open circuit and short circuit. Thus we need to know not just the raw component failure rate, but what proportion of failures are to open and short circuit. Unfortunately there is little information of this kind readily available; MIL-HDBK-217B, a major reference text of failure rates, makes no distinction between open- and short-circuit faults.

Despite the scarcity of data there are some conclusions that can be reached by examining the nature of certain components; for example, it is difficult to envisage a mechanism which will cause a short-circuit failure in a low-power carbon film or metal film resistor so that the predominant failure mode is open circuit. To take care of the remote possibility of a short circuit it is sometimes given a failure probability of 1% of the total failure rate.

Relay contacts also have a predominant failure mode. If they are open and are required to close when the relay coil is energized there is a very small probability that they will not do so because a speck of contamination has landed on one of the contacts and forms an insulating barrier between them, or because the coil is open circuit. In order to diminish the probability of this occurrence twin contacts are provided on a widely used telephone relay so that correct operation is assured even if one of the two contact pairs is non-conducting. An alternative approach is to seal the contacts into a gas-tight enclosure, typically filled with dry nitrogen, so that no dust or contamination can reach the contacts and they cannot oxidize. This is the method adopted in the reed relay which has high reliability but cannot handle such heavy currents as the standard telephone relay nor drive as many contact pairs from one coil.

If a set of relay contacts have already closed and are in contact and the coil is then de-energized it is difficult to postulate a mechanism other than gross mechanical damage to the relay which will prevent the contacts from opening and breaking the circuit. Consequently in analysing failure modes in relay circuits the open-circuit failure rate is often assumed to be some 50 times greater than the short-circuit rate. Here the short-circuit failure rate is associated with the failure of the contacts to open when the current in the relay coil is interrupted.

7.3 Relay tripping circuits

Many alarm and monitoring systems have a variety of sensing elements and are required to operate alarms and in critical situations shut down the plant they control. The sensing elements may be those used for system control or in critical cases a separate set. In both arrangements the alarm action must operate when any one plant variable goes out of range. The corresponding circuit may be a parallel or a series connection. If all of the alarm outputs are connected in parallel they must normally be open circuit and must close to signal a fault. As this mode has a failure rate much higher than that where the relay contacts are normally closed and open to signal a fault it is not generally used. The preferred circuit is that with all the relay contacts in series and with all of them closed, as shown in Figure 7.1.

A fault detected by any unit will open the contact chain and can be made to energize an alarm to take emergency action. The contact opening could be caused either by energizing or de-energizing the relay coil. To minimize dangerous failures the coil is designed to be normally energized and the current is interrupted to indicate a fault. This means that a failure of the

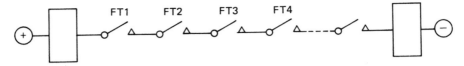

Figure 7.1 Series connection of relay contacts in guard line

relay power supply or an open-circuit fault in the relay coil will be a fail-safe rather than a fail-dangerous event.

7.4 Mechanical fail-safe devices

The same design principle can be applied where objects have to be moved when an alarm is signalled. For example, when a nuclear reactor develops excess power or overheats it must be shut down as soon as possible. This involves inserting control rods made of neutron-absorbing material such as boron into the reactor. One widely used scheme uses vertical channels into which the rods can drop. They are normally held out of the reactor by electromagnets attached to movable rods and any condition calling for shut-down will interrupt the current in the electromagnets so causing the rods to drop into the reactor under gravity. This means that any failure in the power supply or a break in the cabling or the electromagnet winding is a safe failure.

Passenger lifts comprise another group of control systems in which mechanical fail-safe mechanisms find a place. After some early accidents caused by failure of the hoist cable or the drive gearbox public agitation stimulated a search for a safety device which would prevent this kind of accident. All failures in the drive or the cable result in a loss of tension in the connection between the cable and the lift cage. One fail-safe mechanism based upon this fact uses hinged spring-loaded cams attached to the top of the lift which are held back by the tension in the cable. When this tension disappears and the lift begins free fall the cams fly outwards and jam against the lift guides, so stopping the descent of the lift.

7.5 Control system faults

Classifying errors in a control system is more difficult than in an alarm system; in some cases where there is no provision for reversion to manual control any failure may be dangerous. Manned spacecraft are generally controlled by on-board computers, but the crew have on occasions had to take over control. This was possible because critical landing operations are usually carried out in fair weather and in daylight. However, if there is no time to change over to manual control the fault must be classed as dangerous.

The way in which faults in aircraft autopilot controls are classified depends upon the particular phase of the flight in which they occur. For example, in mid flight a fault which deprives the autopilot of power but leaves the aircraft flying a straight course may be comparatively safe as there should be time for the pilot to take over. On the other hand a fault which sends the aircraft into a full-power dive and prevents manual control may cause structural failure and a major accident.

When a so-called 'blind' landing is made in bad weather the pilot has few of the visual indications normally available to show the aircraft's position relative to the runway. The aircraft is guided to the vicinity of the airfield by a radio beacon and then aligned with the runway by further short-range radio beams. A vertical beam gives a guide to the end of the runway and these signals together with the aircraft's terrain clearance measuring radio give all the information needed for an automatic landing. When still several hundred feet above the runway the pilot can decide to abort the landing and either make another attempt to land or divert to another airport. Beyond this point the pilot must hand control over to the automatic control system and no reversion to manual control is possible.

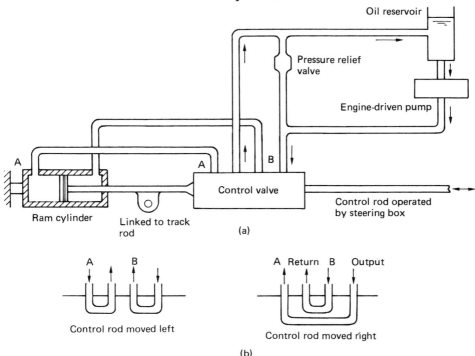

Figure 7.2 Power-assisted steering: (a) layout; (b) flow of hydraulic fluid

84 *Reliability in Instrumentation and Control*

The redundant control system is equipped with a fault detection system which warns the pilot if there is a disagreement between the channels which operate each control surface, so that he or she would normally commit the aircraft to an automatic landing only if there were no fault. The duration of this final phase during which the aircraft must be controlled automatically is about 30 seconds and for this time any failure of the control system must be classed as a dangerous failure.

In today's traffic conditions steering a car requires continual vigilance, and any failure of the steering mechanism would clearly be a dangerous fault. Consequently attempts to reduce the effort needed by the driver using power derived from the engine are designed to be fail safe, so that if the hydraulic power mechanism fails, the driver retains control and reverts to manual steering. In one form of the system a spring-loaded spool-type valve is fitted in the steering rod which connects the steering box to the front wheels, as shown in Figure 7.2. With no torque exerted on the steering wheel, the springs hold the valve centrally and the oil from the pump is returned directly to it. When the driver turns the wheel the springs deflect and allow oil to pass through the valve to a piston and cylinder which are attached to the drag link, so turning the road wheels in the direction the driver desires. When the road wheels have turned as much as the driver wishes, the force in the drag link ceases, the spool valve is returned to the neutral position and the oil supply to the cylinder is cut off. Endstops fitted to the spool valve allow only a small degree of movement between the valve and its housing and transmit the steering force when the hydraulic power assistance fails.

7.6 Circuit fault analysis

In order to estimate the probability of fail-safe and fail-dangerous faults in a circuit it is necessary to examine the consequence of each component sustaining a short- or an open-circuit fault. As an example we take a relay-driving circuit which is driven from an open-circuit collector TTL gate such as the 7403. The circuit is shown in Figure 7.3. The input is normally low, holding T1 and T2 in conduction and the relay energized. To indicate an alarm the input becomes high impedance so that the current in R1 and R2 ceases. T1 and T2 are cut off and the relay current falls to zero. A fail-safe fault will occur when the relay releases with the input held low. This could be caused by the following faults:

```
R2   Open circuit (O/C)    T1   O/C
R3   O/C                   T2   O/C
D1   Short circuit (S/C)   RL   Coil O/C or S/C
                                Contacts O/C
```

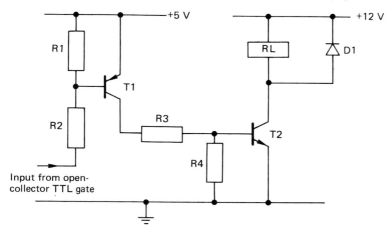

Figure 7.3 Relay-driving circuit

Although there is a small probability of the resistors suffering a short-circuit failure, this is much less than other probabilities which also could cause a fail-safe failure and it has been neglected. A short circuit of R1 and R4 comes into this category.

A fail-dangerous condition would be caused by any fault which would hold the relay contacts closed when the input was open circuit. The obvious faults are T1 or T2 S/C, but R1 or R4 O/C might also give trouble if the ambient temperature were so high as to increase the leakage current in T1 or the input driver. An electrically noisy environment might also cause trouble since R2 would be at a high impedance to earth and would thus be liable to pick up noise voltages which could hold T1 partly conducting. There is also a small probability that the relay contacts will develop a short circuit.

Using the failure rate data in the table below we can estimate the probabilities of fail-safe and fail-dangerous faults:

Component	O/C failure rate (FIT)	S/C Failure rate (FIT)
Transistor	2.5	2.5
Resistor	5	–
Relay coil	15	5
Diode	1.2	0.8
Contacts	10	0.2

1 FIT is a failure rate of 10^{-9} per hour

The probability of a fail-safe fault is then 45.8 FIT, and the probability of a fail-dangerous fault between 5.2 and 15.2 FIT. The lower figure is perhaps a little optimistic since it assumes that neither R1 nor R4 O/C will cause a fault. This is likely to be so if the unit is not subjected to high temperature and is not in a noisy environment. If we cannot be certain of this we must use the figure of 15.2 FIT.

A further possible cause of a safe failure is the collapse of either power supply voltage. If the probability of this occurring is comparable with the component failure rates it must be included in the total. Often, however, several alternative power sources are available with automatic switching and the probability of a power failure may then become much smaller than the other failure probabilities; it can thus be neglected.

A difficulty with any alarm system is that a fail-safe error immediately reveals itself as the plant it is connected to will shut down automatically. A fail-dangerous error on the other hand will not reveal itself until the plant develops a potentially dangerous fault and the expected automatic shutdown does not occur. In order to minimize the likelihood of this it is normal practice to conduct routine maintenance of the system with particular emphasis on checking for fail-dangerous faults. Such maintenance reduces the period during which any particular sensing unit can remain in a fail-dangerous state and so reduces the chance of a serious accident, which can occur only when the sensing unit is in this state and a plant failure also occurs.

Bibliography

Ballard, D. R. (1979) Designing fail-safe microprocessor systems. *Electronics*, 4 January

Celinski, K. (1987) Microcomputer controllers introduce modern technology in fail-safe signalling. *IEE Conferences Publication*, No. 279

Cohen, H. (1984) Space reliability technology: a historical perspective. *IEEE Transactions on Reliability*, **R-33**, No. 1

Curtis, R. and Hogan, E. (1980) *Nuclear Lessons*, Turnstone Press, Wellingborough

Green, A. E. (1982) *High Risk Safety Technology*, Wiley, Chichester

Gregory, R. A. (1984) Specification and assessment of high integrity interlock system. *IEE Colloquium on Computing for Safe Control*

Jervis, M. W. (1984) Control and instrumentation of large nuclear power stations. *IEE Proceedings*, **131**, Pt A, No. 7, p. 481

Klassen, H.B. and van Pappen, J. C. L. (1989) *System Reliability*, Edward Arnold, London

Lisboa, J. J. (1988) Essential information required for an engineering and probabilistic assessment of a reactor protection system. *IEEE Transactions on Nuclear Science*, **NS-35**, No. 1

Smith, D. (1984) Failure to safety in process-control systems. *Electronics & Power*, **30**, March

8 Software reliability

8.1 Comparison with hardware reliability

During the last two decades computers and microprocessors have undergone steady development with particular improvements in operating speeds, storage capacities, variety of instructions and reliability. Also for a given level of performance their size, weight and cost has fallen considerably. In consequence they are being used in ever increasing numbers in instruments and control and instrumentation systems. The improvements in hardware reliability, with the possible use of redundancy, has meant that more and more failures arise not from hardware malfunction but from errors in computer programs. We need thus to address the problem not only of hardware reliability but also of software reliability.

During the last 40 years a great deal of knowledge has been collected about hardware reliability and its assessment, but unfortunately only part of this can be applied to software. We can, for example, overstress semiconductors by operating them at high temperatures to increase the number of failures, with the expectation that the surviving devices will be the more reliable ones. There is unfortunately no similar method of 'overstressing' a computer program to hasten the discovery of faults. Also we expect hardware reliability to decrease as the equipment ages due to wear in mechanical components and ultimately the slow degradation of semiconductors. No such changes are expected with software; once this has been tested fully it does not age and its reliability should remain constant. Where users claim the contrary, it is generally found that the program is being used in a manner for which it was not designed and thus errors are to be expected. Where the error arises from particular values of data it may not occur until long after the program was commissioned.

For example, an early payroll program was written for a small firm having about 200 employees. The programmer thought that he was allowing ample room for growth by allocating 10 bits to encode the employee number, which would cater for a total of just over 1000. However, subsequent expansion and amalgamation raised the number of employees to well over 1000 some years later and the program which had run for years with no trouble suddenly produced errors as some employee numbers could no longer be encoded correctly. By this time the original programmer had long left the firm and the documentation was poor. It consequently took much careful analysis of the program to reveal the error. No testing of the program with typical data at the time of commissioning could have revealed this particular fault.

When analysing the performance of software we can use the same definition of reliability as we use for hardware, namely the probability that the program will operate for a specified duration in a specified environment without a failure. We can also take advantage of redundancy to improve performance and use the same methods of calculation to estimate the overall reliability of a compound program in terms of the reliabilities of the separate sections of the program. The same procedure can also be used to estimate the reliability of a complete system in terms of the separate reliabilities of the hardware and software.

8.2 The distinction between faults and failures

When discussing software reliability it is important to distinguish between faults and failures. A failure occurs when a program fails to produce the output expected (we assume that the hardware is operating correctly). This arises from some error or fault in the program. However, the fault may not cause a failure if the program section in which it lies is not executed. Most control and measurement programs comprise a large number of routines, only some of which are executed during each run of the program, so that a fault may be dormant for some time until the routine which contains it happens to be executed. The large number of different paths through a typical program means that the chance of any particular execution revealing a fault is very small. Some measurements quoted by Musa *et al.* (1987) on a mixture of programs showed that the number of failures per fault varied from 1.878×10^{-7} to 10.6×10^{-7} with an average of 4.2×10^{-7}. The expected number of program runs to detect a fault is thus 2.4×10^{6}. If each run takes 50 milliseconds the time needed is about 33 hours per fault.

The number of failures per fault given above is much less than might have been expected, but some simple examples show figures of a similar order. For example, if some integer data item is allocated 16 bits of storage and all of the 65 536 possible values are equally likely, the chance of any one occurring is about 1.53×10^{-5}. Thus if a program instruction is written as 'Branch if X < 52' instead of 'Branch if X < 50' an error will occur only if the value of X is 50 or 51. The chance of this occurring is then about 3.05×10^{-5}. If all values of X are not equally likely, for example values below 1000 are 50 times less likely than those above 1000, the chance of an error given a random value of X falls to about 6.1×10^{-7}. This is also the likely number of failures per fault.

Equally small values of failures per fault may arise if the failure is only revealed when a particular path through the program is followed. For example, if each run of the program on average includes 20 two-way branches there are about 1.05×10^6 different paths. If only one of these gives a failure and all paths are equally likely the number of failures per fault is the reciprocal of this, about 9.54×10^{-7}. It is also possible that one fault will produce a fractional number of failures generally, if, for example, it occurs in a frequently used subroutine. However, such a fault should be discovered very early in the testing phase and should not appear in any software which has been subjected to a normal testing process and is considered ready for release.

In off-line computer operations such as data processing the time required for the execution of a program segment is not usually critical: the important factor is the total time needed to process a suite of programs such as, for example, those involved in payroll calculations. The situation may be different in on-line operations such as vehicle or process control. Here the calculations involved in a control loop must be performed in a certain time, otherwise the system becomes unstable. Thus there is an additional failure mode which must be considered; the right answer is no use if it is not available in time. This point may be a restriction in the use of software redundancy schemes which involves extra computation to mask a fault.

8.3 Typical failure intensities

The performance of a program can be specified in a number of ways; reliability is usually most useful when a run or mission of a specified duration is involved. However, as with hardware this depends upon specifying a particular duration and if this is altered a different value of reliability occurs. As with hardware we can specify a failure rate which does not depend on run time, usually called failure intensity in software. If, for example, this is 0.001 failures per hour we would expect the program on average to operate for 1000 hours without a failure. The probability of this is $\exp(-0.001 \times 1000) = 0.368$, and consequently the reliability of the software for a 1000-hour operating period is also 0.368.

Although the failure intensity is an important property of a program it is dependent on the environment, in particular the rate at which instructions are executed. If we measure an average failure intensity of 0.002 failures per hour for a particular microprocessor system with a clock speed of 2 MHz, we would expect a failure intensity of 0.004 failures per hour if the clock speed were raised to 4 MHz. The program failure statistic which does not alter with the environment is the average number of faults in a given section of program (usually 1000 instructions). This is usually classified as the fault density.

Computer folklore has it that many commercial programs contain about one fault per 1000 instructions when first released. Measurements reported (Musa 1987) for programs of about 100 000 lines of source code show variations from 1.4 to 3.9 faults per 1000 source lines when first operational and so support this assertion. The fault density is of course much greater, possibly by a factor of a hundred or so, when the program is first compiled or assembled. Many of these faults will be detected by the compiler and others will be fairly obvious, so that the initial debugging effort will rapidly reduce the number of faults. However, since only a failure will reveal a fault, each time a fault is removed, the average CPU time required to reveal the next fault is increased. Thus the effort needed to discover and correct a fault steadily increases and for much commercial software the cost of trying to eliminate all faults becomes too much. It would also take a long time.

If the software is released when nearly all of the faults have been removed to say 100 customers, all of whom use it regularly, the remaining faults are likely to be discovered much more quickly than if the only testing is undertaken by the program suppliers. Users are often prepared to test new programs under working conditions in return for early experience of their use.

8.4 High-reliability software

Although the release of new software which cannot be guaranteed free from errors is the usual practice for commercial applications, it is not acceptable for software designed for safety-critical situations, such as aircraft controls and the monitoring of nuclear power reactors. The cost of a failure in terms of money or human lives is so great in these cases that extreme efforts are required to produce error-free programs. For example, the error rate quoted for the 'fly-by-wire' controls of the A320 airbus is 10^{-9} per hour. For earlier passenger aircraft error rates of 10^{-9} per mission were proposed, a mission being a flight of 1 to 10 hours' duration. In order to construct software to this standard it is necessary to specify the requirement in very precise terms, perhaps using a language specially designed for the purpose. The program must be structured so that it can be divided into fairly small modules, each of which can be tested separately and has a well-defined interface to the rest of the program. Languages such as Coral and Ada have been designed for applications of this kind. The attainment of such low error rates usually requires in addition some degree of redundancy.

The cost of developing large, very reliable programs may inhibit the building of some systems in which safety is a critical factor. For example, Uhrig (Uhrig 89) estimates that a program to automate the control, monitoring and safety functions of a single advanced nuclear power plant would contain towards a million lines of code. This is much greater than the program used for a current LWR, but rather less than that for a Space Shuttle mission. Using the traditional error rate of 1 bug per 1000 lines of code there will be 1000 bugs to be removed. The average cost of a fully tested program of this sort is $60 to $100 per line. Thus the cost of software alone could be as much as $100 million. This would not be acceptable unless it could be shared over many reactors.

Some reduction in the cost is possible by using computer-aided software engineering (CASE) in which a computer helps to write the program, but even then it is still very expensive.

8.5 Models of reliability growth

It was noted in Section 8.3 that the most fundamental measure of program reliability is the fault density, that is the number of faults per 1000 instructions. This is independent of the speed of the processor used. However, engineers are used to measuring and specifying reliability in terms of time, and failure rates of systems are specified as a function of time rather than, for example, per million components. Thus there is an advantage in using execution time as the variable in measuring software reliability. This is particularly important when we wish to combine software and hardware reliability to assess overall system reliability. When considering software failure rates we can easily allow for the effect of changing the processor, so changing the number of instructions executed per second, and for executing only part of the program.

We have seen that almost all software has far too many faults when initially compiled, and much effort must then be devoted to finding and removing nearly all of them. This effort may be up to half of the total resources allotted to generating the program. The management of the task is made much easier if some estimate can be made of the rate at which faults will be found and removed, and of the consequent increase in reliability as a function of time. Most software has a delivery date by which it is expected to be reliable enough for release, and it is useful to compare this with the projected growth in reliability. If the reliability is not expected to grow rapidly enough more resources can be allocated to the task to bring forward the projected release date.

This process of estimation requires a model of the reliability growth, usually as a function of execution time. Several of these have been proposed, but a simple one which fits many situations quite well is the basic run-time model. This predicates a linear relation between the number of failures and the failure intensity, of the form

$$\lambda(n) = \lambda_0(1 - n/n_0) \text{ failures per hour} \tag{8.1}$$

Here λ_0 is the initial failure intensity, n_0 the total number of failures in infinite time and n the number of failures since execution started. In this

92 Reliability in Instrumentation and Control

equation n must be an integer so that the graph of failure intensity against the number of failures must be stepped. However, if the program is fairly large with many expected failures we can approximate to the situation by regarding n as a continuous variable. The failure intensity then becomes the time derivative of the number of failures or dn/dt. This leads to the differential equation

$$dn/dt = \lambda_0 (1 - n/n_0) \qquad (8.2)$$

The solution of this is

$$n = n_0 [1 - \exp(-\lambda_0 t/n_0]) \qquad (8.3)$$

Thus the number of failures is an exponential function of time and the time between successive failures is a steadily increasing interval. Figure 8.1 shows the relation between the failure intensity and the number of failures and Figure 8.2 the relation between the number of failures and the execution time.

We can combine Equation (8.1) and (8.3) to give a relation between the failure intensity and time:

$$\lambda(t) = \lambda_0 \exp(-\lambda_0 t/n_0) \qquad (8.4)$$

We can transform this equation to give the time as a function of failure intensity:

$$t = (n_0/\lambda_0) \log_e (\lambda_0/\lambda) \qquad (8.5)$$

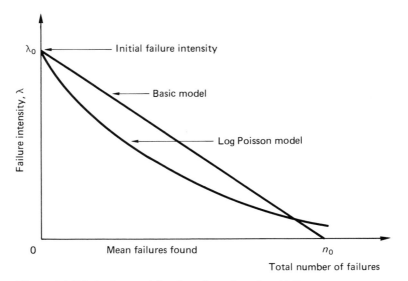

Figure 8.1 *Relation between failure intensity and number of failures*

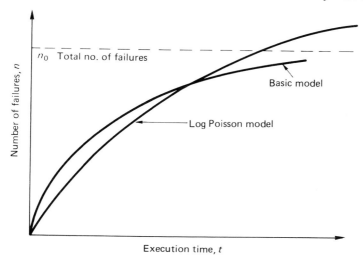

Figure 8.2 *Relation between number of failures and execution time*

A more recently developed model is the logarithmic Poisson model which has an exponential relation between failure intensity and the number of failures of the form

$$\lambda(n) = \lambda_0 \exp(-\theta n) \tag{8.6}$$

Treating n as a continuous variable gives the corresponding relation between n and time t as

$$n = (1/\theta)\log_e(\lambda_0 \theta t + 1) \tag{8.7}$$

Here θ is the failure decay intensity parameter, usually given the value of 0.02. This means that the failure intensity decreases by a factor of 0.98 after each failure. In the basic execution time model the failure intensity decreases by a fixed amount λ_0/n_0 after each failure.

Equation (8.7) can be transformed to give time as a function of n:

$$t = [\exp(\theta n) - 1]/\lambda_0 \theta \tag{8.8}$$

For example, we consider a program which would experience 150 failures in an infinite time. If it has already had 40 failures, each of which has caused remedial action, and the initial failure rate was 6 per hour, the present failure rate is given by the basic run time model as

$6 \times (1 - 40/150) = 4.4$ failures per hour using Equation (8.1). After 20 failures the failure rate will be

$6 \times (1 - 20/150) = 5.2$ failures per hour

The time required to reduce the failure intensity to 4.4 per hour is given by Equation (8.5) as

$t = (150/6)\log_e(6/4.4) = 7.75$ hours

Using the log Poisson model as in Equation (8.6) the failure rate after 40 failures is

$6 \times \exp(-0.02 \times 40) = 2.696$ failures per hour

After 20 failures the failure rate would be

$6 \times \exp(-0.02 \times 20) = 4.02$ failures per hour

These calculations using the log Poisson model assume a value of 0.02 for the failure intensity decay parameter θ and do not agree very well with those using the basic run-time model. If the parameter θ is decreased to 0.01 the agreement is much better, the failure rates after 40 and 20 failures then becoming 4.02 and 4.91 failures per hour compared with 4.04 and 5.2 failures per hour using the basic model.

The time needed to deal with 40 faults is given for the log Poisson model by

$t = [\exp(0.02 \times 40) - 1]/6 \times 0.02 = 10.2$ hours
using a value of 0.02 for θ.

One small adjustment which needs to be made to calculations such as those above is a consequence of the imperfect process of 'debugging' or error removal. It is not realistic to expect that the programmers engaged in error removal will perform this task perfectly; experience shows that a small number of other faults will be introduced during the operation. Figures for work on programs varying from 22 000 to 100 000 source lines in size show that the number of faults introduced per 100 faults corrected were between 1 and 6 (Musa *et al.* 1987). Thus the number of faults removed must be multiplied by a fault reduction factor B in the range 0.94 to 0.99 to allow for the new faults introduced in the process.

8.6 Estimating the number of faults

In order to monitor the progress of debugging and to estimate the size of the task it would be useful to have some estimate of the number of faults in the program being tackled. One method of doing this is called 'seeding' and is based upon a technique used to estimate the number of fish of a certain species living in a large pond. The pond is seeded by introducing a number of similar fish which have been tagged into the pond. After a suitable interval to allow the tagged fish to become evenly distributed throughout the pond, a sample of fish are removed from the pond and the number of tagged and unmarked fish is counted. If there are 10 tagged fish and 80 unmarked fish of the species being investigated in the sample and 150 tagged fish were added to the pond, we conclude that the population of the pond is 80 × 150/10 = 1200 fish.

In the case of a computer program the seeding process consists of adding errors to the program before the debugging activity is started. The debugging team is not aware of this and reports all faults found. The supervisor then counts the number of deliberate faults found (say 10) and the number of genuine faults found (say 50). If 40 faults were introduced deliberately the estimated total number of faults is 50 × 40/10 = 200.

Although this technique has proved useful, the general experience is that it often given an underestimate. The reason given for this is that the deliberate faults are often easier to find than the original faults, so more of them are discovered than expected. The method is based on the assumption that all errors are equally likely to be found during a particular session of debugging; if the deliberate faults are more likely to be found than the genuine faults, the estimate will be biased.

8.7 Examples of software faults

Nearly all control and instrumentation systems with embedded computers operate on-line, so that the results of a failure are immediately apparent and may be dangerous. Despite exhaustive testing errors still occur; for example, in the American space programme an unmanned probe sent to Venus veered far off course when a slight correction was applied because of a small error in the Fortran guidance program. A period had been written in a DO statement instead of a comma (Bell *et al.* 1987). Another example was reported from an American hospital involving a linear electron accelerator used for X-ray or electron radiation treatment. A fault in the program used to feed-in the characteristics of a particular patient and the dose needed apparently resulted in a massive overdose of radiation which caused the death of two patients before the error was recognized. The overdose was caused because a metal shutter which should have been moved into the path of the beam when changing from X-ray to electron radiation in order to reduce the beam intensity was not operated. Industrial robots are usually operated in a fenced-off area to prevent them coming into contact with people working nearby. However, for maintenance and setting up the robot engineers need to move into the enclosed area and several fatal accidents and injuries have been caused when robots suddenly went out of control due to software errors.

8.8 Structured programming

Many techniques have been proposed to improve program reliability and to make programs easier to write and check. One of these which has gained almost universal acceptance is structured programming (Dahl *et al.* 1972). This has been defined in several ways, but essentially it involves dividing the program into modules which can be compiled and tested separately, and which should have only one entry and one exit. Consequently GOTO statements should be avoided and only three types of program block are required. The first is the sequence in which statements are executed in turn, without branching. The second is the loop in which a set of statements are executed while a condition is satisfied (while . . . do). The third is the selection in which one set of statements is executed if a condition is satisfied, if not another set is executed (if . . . then . . . else . . .). Flowcharts for these constructs are shown in Figure 8.3. Some languages allow for a number of alternatives in the selection process but the same result can be obtained by repeated use of 'if . . . then . . . else . . . '.

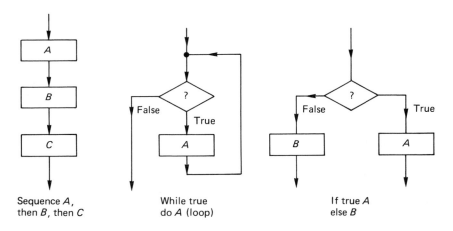

Figure 8.3 *Flowcharts for structured programming*

The size of the modules is a matter of some compromise; if they are too large they become too complex to be easily understood and checked and if they are too small the number required for any significant project becomes excessive. For many purposes a module of 40-50 lines seems appropriate.

One rather extreme view based upon experimental results from psychology is that seven lines or less is the ideal size, and in no circumstances more than nine. This is because we do not seem able to retain in short-term memory more than about seven items and study them as a related set of objects (Miller 1956). However, so many modules would be required for anything but a trivial program that this option is not a practicable one.

One of the early approaches to the design of structured programs was the Michael Jackson or data structure design method (Jackson 75). The basic feature of this is that the structure of the program should correspond to the structure of the files that the program is intended to process. The data is analysed to produce a data structure diagram in which the components of each data item are placed in a box below it and connected to it by lines. Thus in designing a program to produce a report the item 'line' would be below the item 'page' which in turn would be below 'report'. The program structure can then be produced by replacing each item by a module which processes it. Thus 'line' is replaced by a module above which generates a page. Modules on the same level are normally processed as a sequence, but two of them may be alternatives and they are marked to denote this. The modules defined in this way are all of the three types mentioned above, sequences, loops or selections.

A later development which has been promoted by the National Computing Centre is structured systems analysis and design methodology (SSADM) which is now claimed to be the most widely used structured method in Europe. It is also being submitted as the basis for a European software development methodology.

Experience has shown that the use of structured methods in programming makes for fewer errors, needs less time in writing and testing programs and makes the subsequent maintenance and enhancement of programs much easier. Its advantages are particularly evident in large projects where many programmers are each working on a small part of the program and need precisely defined interfaces between their various segments. It is often convenient to make each module a subroutine or procedure; this ensures that any variables used in the subroutine cannot be accessed from outside the subroutine unless they have been declared as 'global'. We can then use the same variable names in a number of different modules without confusion. This is particularly useful in developing large programs using many programmers who do not then need to keep a constantly changing list of all variable names used in the project.

Since structured programming was first introduced many versions of it have been proposed some of which depend upon the use of particular languages. Early high-level languages such as Fortran were not designed with structured programming in mind. Later languages such as Algol and Pascal were specifically given features which simplified the division of the program into independent modules with a well-defined start and finish, as required for structured programming.

8.9 Languages for high-integrity programs

Experience with early high-level languages revealed that they contained a number of features which were liable to cause errors, unless extreme care was taken in writing and testing programs. For example, their compilers generally did not check the value of the index when an element of an array was accessed, to ensure that it was within permitted bounds. The elements of arrays and other variables are usually held in contiguous areas of storage, so that if the program tries to access an array element with an out-of-bounds index it will access an element of another array, or a single variable. For example, if arrays P with 100 elements and Q with 50 elements are stored in locations 500-599 and 600-649 (decimal) and an incorrect calculation gives the index value to access P as 120, the program will access array Q. The compiler will look up the base address of P(500) and add 120 to it. It will then access location 620 which contains the array element Q(20). The way to avoid errors of this sort is to make the program check that the index for any access to an array must lie within the bounds set by the amount of storage allocated to it, typically by a DIMENSION statement. In this example the indices for P and Q must not exceed 100 and 50 respectively. If an attempt was made to access P(120) the program would stop and an error message such as 'Array index out of bounds' would be displayed. This gives a good indication of the fault and its location, but in a critical real-time application such as aircraft control a program termination of this sort would not be acceptable. In such a case rigorous testing must be undertaken to ensure that this fault does not occur.

In early versions of Fortran it was not necessary to declare variables before using them and a new variable name could be introduced into the program at any point. This could cause errors if a variable were misspelt. For example, if we were using a variable CURRT and wrote an expression to increment it as

CURRT = CURT + 1

the result could be indeterminate since we have introduced a new variable CURT without assigning it a value. Depending on the compiler used, the value read from the location CURT might be zero, if storage were initially cleared, or a random value if no attempt was made to alter the value left by a previous run.

Later languages required a list of all variables used to be given at the beginning of the program, with a statement of their type such as Integer, Real, Array. The error caused by misspelling would be picked up at compile time with a display such as 'Undeclared variable CURT'. Some compilers also require all variables to have a value assigned to them before they can be used in a calculation. This feature avoids the possible introduction of random values into a calculation when variables are not initialized.

Some languages have been developed specifically for high-integrity applications, notably Ada which was sponsored by the American Department of Defense and has also been specified in the UK for defence applications. It was adopted by the US DoD in 1979 and the current ANSI standard version of the language was issued in 1983. It is also being proposed as an ISO standard. While it addresses many of the problems which can arise when using earlier languages it is large and complex. Consequently it is difficult to ensure that all Ada compilers produce the same object code from a given source program, since the language contains some ambiguities.

Ada was little used initially for other than defence applications but the development of reliable compilers which were tested against the ANSI standard has encouraged some commercial applications, for example railway control and the monitoring and control of nuclear power stations.

Another language which has been developed at RSRE specifically for high-integrity applications is Newspeak (Sennett 1989). This addresses the problem of limited word and register sizes by incorporating into the compiler facilities for checking that arithmetic operations cannot cause overflow. This is done when declaring each variable by stating the range of values which it can take; thus, for example, a variable of type byte must lie within the range 0-255 (decimal). This information allows the compiler to check whether arithmetic operations will cause overflow or other errors by evaluating them with extreme values of the variables. Any expression which could cause overflow is rejected by the compiler. This arrangement eliminates run-time errors which could be disastrous in a system controlling an aircraft or a nuclear reactor, which is required to operate for long periods without a failure.

Another possible source of error which is eliminated by Newspeak is the possibility of confusion caused by collecting data in different units. Almost all of the input signals handled by a computer embedded in a control or instrumentation system represent some physical quantity and are consequently meaningless unless the units of measurement are known. Newspeak allows the unit to be associated with each data item and thus errors due to combining data items expressed in different units can be avoided, since only variables measured in the same units can be included in the same calculation. If this condition is not satisfied, the compiler will reject the program. Mixed units could occur, for example in maritime navigation where chart depths are shown in metres and sonar depth sounders may indicate feet or fathoms.

8.10 Job specification languages

Experience has shown that a fruitful area for introducing errors is in the specification of the task to be performed. This is generally written in English with all of the scope for ambiguity inevitable in the use of a natural language. In an attempt to avoid this formal languages have been developed which can be used to specify the task required in a form which avoids all ambiguity. One of the advantages which may ultimately follow is the use of a translation program which could convert the specification into a working program. However, at present this is practicable only for comparatively trivial programs. Examples of job specification languages are Z developed at RSRE (Sennett 1989) and Gypsy developed at the University of Texas at Austin (Good *et al.* 1986).

Although the conversion from specification to binary program is a more reliable process when a job specification language is used, there is still the problem that most users of high-integrity software will only understand a specification written in natural language. Thus there are likely to be ambiguities which have to be resolved; these will occur in the translation from specification into job specification language, rather than into some high-level computer language. Since this involves expressing the same information in a different form, it is likely to attract fewer errors than translating directly into a high-level language, which needs a substantial change and expansion of the original specification.

8.11 Failure-tolerant systems

In previous sections we have discussed ways of generating error-free programs. This can be a slow and expensive process, and for some purposes, particularly when very fast response is not needed, it may be simpler to accept that the program may contain a few errors and design systems which can survive such faults. One method uses what are called recovery blocks.

The procedure is to insert a check on the validity of the output of each small program segment and include frequent dumps of the system state. As soon as an error is detected the program is restarted from the previous dump but an alternative program module is used which should avoid repeating the software error. This involves a considerable increase in the program size above the basic non-redundant version since extra modules are needed both for checking and also for providing alternative ways of performing the task of each basic module. The extra processing delay may be unacceptable for control systems since these need not only correct data but also timely data, so that if calculations take too long they are unacceptable and will result in an unstable system. This rules out the method for systems which require a rapid response.

An alternative method which can be much quicker in operation relies on redundancy; multiple processors, each handling the same data, are used and after each small transaction the outputs are compared. If there is any discrepancy some decision process such as majority voting is used to decide on the most likely result which is used for further processing. The benefit of a properly designed redundancy scheme in minimizing the consequence of hardware faults is well established and potentially the same improvement in reliability can be obtained in relation to software faults. However, the two situations are not exactly comparable since great care is taken to avoid common mode hardware faults, so that a fault in one redundant subsystem has no effect on the probability of a fault in another similar but independent subsystem. This is not the case if all subsystems use the same program, since a fault in this will affect all subsystems equally and no advantage will be obtained from the redundancy.

The way to avoid this is to use different programs for all of the redundant subsystems, so that a fault in one will not necessarily imply a fault in any other. This is clearly an expensive solution since it will be effective only if each team developing software for a particular processor has no knowledge of the work of the other teams. It can be justified only if the application requires a very low failure rate; for example, in the A310 Airbus slat and flap control system two independently developed versions of the software are used. Also, in the American Space Shuttle a single backup computer runs in parallel with four primary computers. All critical functions can be performed by either of two separate completely independent programs, one in the primary computers and the other in the backup computer (Anderson 1979).

Although the use of several independently produced programs can help in constructing fault-tolerant systems, there is some evidence that the programming teams, having had similar training and sharing many attitudes, tend to introduce similar kinds of fault so that it is very difficult to ensure that the resulting programs are truly independent. Also it is unlikely that independent programs will all execute in the same time, so that there will be some delay while all results appear before the decision-making process can be invoked. The task of comparing the outputs of the various modules which are executing the same task is simple only if integer arithmetic is used. However, it is necessary to use floating point numbers for some calculations and different algorithms may use different round-off processes, even when performing the same task. Thus it is necessary to use some approximate comparison process in which numbers are taken to be equal if they differ by only a small prescribed amount.

8.12 Program testing

However much care is devoted to the development of a new program the user will expect the program to undergo testing to demonstrate that it performs the task specified for it. Such acceptance tests are the last operation in the process of software generation but good programming practice insists that checking and testing should be involved at all stages of program development, starting with the specification and flowchart. When coding begins each module should be tested as thoroughly as possible before attempting to run several modules together. The argument for this process is that the earlier any error is discovered and corrected, the less will be the time and effort involved.

A similar argument applies to the finding and correction of hardware faults. These are much easier to detect when testing individual modules than when testing a complete system.

Bibliography

Anderson, R. T. (1979) *Proving Programs Correct*. Wiley, New York
Anderson, T. and Randell, B. (eds) (1979) *Computing System Reliability*, Cambridge University Press
Asami, K. *et al.* (1988) Super-high reliability fault-tolerant system. *IEEE Transaction as Industrial Electronics*, **IE-33**, 148
Baber, R. L. (1991) *Error-free Software*, Wiley, Chichester
Bell, D. *et al.* (1987) *Software Engineering: A Programming Approach*, Prentice-Hall, Englewood Cliffs, NJ
Bennett, P. A. (1989) The march towards standards in safety related systems. *IEE Conference Proceedings*, No. 314, p. 106
Bloomfield, R. E., (1986) The application of formal methods to the assessment of high integrity software. *IEEE Transactions on Software Engineering*, **SE-12**, 988
Brozendale, J. (1989) A framework for achieving safety-integrity in software. *IEE Conference Proceedings*, No. 314, p. 13
Cullyer, W. J. and Wise, J. W. (1989) Formal methods for railway signalling, *IEE Conference Proceedings*, No. 314, p. 86
Dahl, O. J. *et al.* (1972) *Structured Programming*, Academic Press, London
Dijkstra, E. W. (1990) *Formal Development of Programs and Proofs*, Addison-Wesley, Reading, MA
Dunn, R. H. (1984) *Software Defect Removal*, McGraw-Hill, New York
Evans, M. W. and Marcinniak, J. (1987) *Software Quality Assurance and Management*, Wiley, Chichester
Good, D. I. *et al.* (1986) Report of Gipsy 2.05. Institute for Computing Science, University of Texas at Austin
Hamilton, M. H. (1986) Zero-defect software: the elusive goal. *IEEE Spectrum*, **48**
Hill, J. V. (1988) The development of high reliability software. *IEE Conference Proceedings*, No. 290, p. 169
Hill, J. V. *et al.* (1989) Safety critical software in control systems – a project review. *IEE Conference Proceedings*, No. 314, p. 106
Ince, D. (ed.) (1986) *Software Engineering – the Decade of Change*, Peter Peregrinus, London
Jackson, M. (1975) *Principles of Program Design*, Academic Press, London
Kershaw, J. (1989) Dependable systems using 'VIPER'. *IEE Conference Proceedings*, No. 314, 23
Koch, H. and Kubat, P. (1983) Optimal release time of computer software. *IEEE Transaction Software Engineering*. **SE-9**, 323
Markham, K. C. and Milliken, R. A. (1989) Software fault tolerance for a flight control system. *IEE Conference Proceedings*, No. 314, p. 18

Miller, G. A. (1956) The magical number seven, plus or minus two; limits on our capacity for processing information. *The Psychological Review*, **63**, No. 2, 81

Musa, J. et al. (1987) *Software Reliability: Measurement, Production, Application*, McGraw-Hill, New York

Myers, G. J. (1979) *The Art of Software Testing*, Wiley, New York

Ould, M. A. and Unwin, C. (1986) *Testing in software Development*, Cambridge University Press

Rodd, M. G. and Zhao, G. F. (1989) Assessment of real-time software – a review. *IEE Conference Proceedings*, No. 314, p. 92

Sefton, B. (1989) Safety related instrument systems for the process industries. *IEE Conference Proceedings*, No. 314, p. 41

Sennett, C. (1989) *High-Integrity Software*, Pitman, London

Shooman, M. L. (1983) *Software Engineering; Design, Reliability and Management*, McGraw-Hill, New York

Shooman, M. L. (1984) Software reliability; A historic perspective. *IEEE Transactions on Reliability*, **R-33**, 48

Short, R. C. (1987) Software validation for railway signalling train control systems. *IEE Conference Proceedings*, No. 279, p. 315

Soi, I. M. and Aggarwal, K. K. (1980) On reliable software development for microprocessors. *Microelectronics & Reliability*, **20**, 273

Thomas, N. C. and Reeves, H. L. (1980) Experience from quality assurance in nuclear power plant protection system software validation. *IEEE Transactions on Nuclear Science*, **NS-27**, 899

Trachtenberg, M. (1985) The linear software reliability model and uniform testing. *IEEE Transactions on Reliability*, **R-34**, 8

Uhrig, R. E. (1989) Automation and control of next-generation nuclear power plants. *Nuclear Technology*, **88**, 157

Ward, M. (1990) *Software that Works*, Academic Press, London

9 Data transmission

9.1 The need for data transmission

As the techniques of instrumentation and control have developed, their size and complexity have increased in a spectacular fashion; in particular they involve the transmission of data and control signals over ever increasing distances.

For many years signals were confined to a single chemical plant, refinery, boiler house or power station. The advent of reliable world-wide communication has enabled control to be exercised over much greater distances; for example, astronomers can now collect data from telescopes installed on a mountain top on the other side of the world and select the area in the sky at which they point. Also the manoeuvres of satellites and spacecraft can now be controlled from ground-based laboratories which may be over 20 000 miles away.

In all of these situations it is essential to ensure that the data and control signals are transmitted accurately, despite possible variations in channel transmission and the intrusion of noise and interference. Errors in transmission can occur due to a complete failure of a link, or momentarily due to a noise pulse or an increase in channel attenuation. The relative likelihood of these two events depends on the means of transmission and will determine the most effective method of reducing errors. For example, a 50-mile (80 km) telephone circuit is very unlikely to fail completely but will probably pass through several exchanges. In each of these engineers will be working on altering existing circuits and connecting up new ones and may temporarily short-circuit the wanted pair or inject unwanted voltages on to it. Thus for reliable transmission we need a mechanism which will cope with short interruptions and noise spikes. In contrast a submarine optical fibre cable has a negligible chance of picking up any interference, but may suffer rare catastrophic failures and so is usually provided with a degree of hardware redundancy. Measures to improve reliability are more effective with digitally coded systems, but analogue coding still has a place in instrumentation and control systems.

9.2 Analogue transmission

Many forms of transducer have analogue outputs, as do most pointer indicators and chart recorders. Thus in the interest of economy and simplicity the use of analogue signals where possible seems a natural choice since it minimizes the need for analogue-to-digital (A–D) and digital-to-analogue (D–A) conversion. However, it decreases the ways in which the effect of noise and circuit interruptions can be minimized. Comparatively short links, such as would be found in a power station or chemical plant, should be very unlikely to suffer a catastrophic failure and can in any case easily be duplicated. They often carry signals with a low bandwidth; for example, the temperature of any appreciable mass of liquid or solid cannot change very quickly, so that much wideband noise and interference can be eliminated by passing the signals through a low-pass filter. Apart from single-frequency interference such as that induced from power lines, most interference is wideband and thus much of it can be eliminated by using a low-pass filter in series with the system input whose cut-off frequency is a little above the maximum frequency component of the input signal.

A common problem in an electrically noisy environment such as a power station or a factory using electrical welding is the pickup of common mode noise voltages. This is best combated by using screened twin-wire cables, with differential driving and receiving circuits. If the instantaneous voltages on the two lines are V_1 and V_2 with respect to earth, the common mode or longitudinal voltage is defined as $(V_1+V_2)/2$ and the transverse or signal voltage is (V_1-V_2). By using long-tailed pair circuits with, if necessary, common mode feedback we can produce line receiving amplifiers which can detect a small signal voltage in the presence of a much larger common mode voltage. Recently the cost of fibre optic cables has fallen to a level where they can be considered for use in control and instrumentation systems which work in a hostile electrical environment. Recent work under a European Esprit project is intended to develop optical fibre systems which are economic for short runs of 1-10 metres (Parker 1991). They have the advantage that they are completely immune from electrical interference and so the circuits into which they are coupled do not need any filtering.

9.3 Analogue transducers

A survey of the transducers used in typical instrumental and control systems reveals that the majority of these deliver an analogue output. Amongst these transducers are almost all temperature sensors such as thermocouples and thermistors, strain gauges, inductive and capacitive proximity sensors, potentiometric position sensors and tachometers used to measure angular speed.

The reliability of the transducer data can be improved by protecting the transducer and its wiring as far as possible from electrical interference and the effect of any hostile environment. Beyond this, additional improvement can be obtained by periodic calibration tests and extra circuits to detect open- and short-circuit faults, and by providing redundant transducers and cabling.

9.4 Data validation

The reliability of a measuring system can be improved by checking the validity of the input data before accepting it into the system and rejecting any which is incorrect. With a purely analogue system it is possible to fit voltage-sensing circuits which will detect any signal outside a prescribed voltage band. This will allow any out-of-range signal to be rejected, but it cannot detect any smaller error.

If a computer is incorporated into the system, much more precise checks may be performed. These are conducted on the data within the computer so that they will cover the transducer, the ADC and their interconnections. One scheme is to calculate the rate of change of data. In most applications the mechanical features of the system provide an inherent limit to the rate of change of the input; for example, the thermal inertia of a component and the maximum rate at which heat can be added or removed will limit the rate of change of temperature. Also any motor driving part of a control system will have a maximum speed and thus any displacement transducer measuring the position of the part will have a limit placed on the rate if change of its output. In these cases the computer can be programmed to reject any input which is changing too quickly.

In many control and instrumentation systems multiple transducers are located comparatively near to one another so as to provide detailed information about the plant. This is commonly the case with temperature measurements. In these conditions it is possible to compare the output of each transducer with the outputs of adjacent transducers. Operating experience will indicate the likely disparity between adjacent transducer outputs, and any value significantly above this can be regarded as an error.

Unless it is possible to send calibrating signals from the transducers periodically, there is no simple mechanism for detecting a slow drift in the attenuation of the transmission lines between transducers and the computer, when using analogue transmission. The situation is quite different, however, if digital coding is used.

9.5 Digitally coded signals

With digital coding, signal amplitude can vary widely without causing data errors. For quite short links CMOS logic can tolerate some 20% signal attenuation without error. Longer links are usually operated over a balanced pair of lines, and these can tolerate a much greater attenuation. By using special line driver packages a differential signal of some 4-8 V can be transmitted and the line receivers typically have sensitivities of ±25 mV. Thus the signals could suffer an attenuation by a factor of 100 and the received voltage would still be well above the threshold. A further advantage of digital coding is that when links are connected in series, provided that the signal is restored to its original value at the end of each link, there is no accumulation of noise or interference along the link. This is not true for analogue circuits, where the noise in each link is added to the signal, and the signal-to-noise ratio is less at the output of each link than at the end of the previous link. Thus there is a limit to the number of links which can be used.

9.6 Error detection and error correction

A particular advantage of digital coding is the opportunity it provides for error detection and error correction. The simplest scheme gives only error detection and this is mainly beneficial where the data link can transmit in both directions. The transmitted data is divided into blocks and at the end of the block a check sum is added. This can be derived from the data in several ways; a simple one is to add all the data items together and use only the 8 or 16 least significant bits of the sum. The check sum is computed by the data receiver and compared with the transmitted sum. If they agree the block data is accepted, otherwise a retransmission of the block is requested. In this scheme there is an optimum size of block, depending on the mean error rate. If the block is too small, the check sum takes up a large proportion of the transmission time, so reducing the throughput of data. If the block is too large, so many of the blocks will contain an error and need repeating that the throughput will also be low.

A simple form of error-detecting code used when sending 8 bit characters involves adding an extra 'parity' bit to each character. With odd parity this is chosen to make the total number of ones in the 9 bits an odd number; for even parity the number of ones is an even number. The parity is checked at the receiver and unless there is agreement the character is rejected, often being replaced by an asterisk.

This scheme will detect a single error, but if 2 bits have been inverted the parity will be unchanged and no error will be detected. In most cases the probability of two or more errors in a single character is very small, so the addition of a parity bit is an efficient way of error detection. For example, if we assume a circuit over which 1 bit in 10^4 is inverted, the probability of bit errors in a 9-bit character is

no error	$\sim 1 - 9 \times 10^{-4}$
one error	9×10^{-4}
two errors	3.6×10^{-7}
three errors	8.4×10^{-11}

The figure for no error is obtained accurately by taking the sum of the probabilities of one, two, three, etc., errors and subtracting this from one. The probability of a single error is so much greater than all of the other probabilities that we can neglect them in calculating the chance of no error.

The figures above show that the chance of two errors in a character (which would not be detected) is 2500 times less than the chance of one error. Thus the rate of undetected error is reduced from 9×10^{-4} per character to 3.6×10^{-8} by introducing the parity bit. If arrangements are made to retransmit any group of characters which contains a parity failure we can reduce the error rate from 1 character in 1111 to 1 in 2.78×10^7.

Where the circuit has a lower error rate the data may be assembled into larger blocks, each with a check sum following it. As before at the receiver the check sum is also calculated and if it agrees with the sum transmitted the block is accepted as correct.

For critical applications there is a need for error correction. The simplest scheme is one which will correct any single error in a block; this requires a number of check bits to be added to the block. The smallest useful group comprises 4 data bits and 3 check bits. The code is called a Hamming code (Hamming 1950) after its inventor and the group size is of the form $2^n - 1$. This gives the maximum number of data bits which can have a single error corrected by the prescribed number of check bits. Thus the next group comprises 11 data bits and 4 check bits. The principle of error-correcting coding is that only a limited number of the possible bit combinations are allowed. In the 4 + 3 code, the 7 bits can provide $2^7 = 128$ combinations. The 4 data bits can, however, only give $2^4 = 16$ combinations. For each data word the 3 check bits are unique. Each of the complete 7-bit groups has a minimum 'Hamming distance' from all others; in this case the distance is 3. This means that 3 bits have to be changed to alter one valid group into another. Thus the group 0110101 is distance 3 from the group 0101100 since the third, fourth and seventh bits have to be altered to change one group to the other.

If the minimum distance between groups is d_{min} and the code can detect a maximum of t_D errors and correct a maximum of t_C errors we have the following relations (Wiggert 1988):

$d_{min} \geq 2t_C + 1 \quad d_{min} \geq t_D + 1$

The efficiency can be measured by the ratio of data bits to the total block size. For the 4 + 3 code this is only 57%, increasing to 73% for the 11 + 4 and 84% for the 26 + 5 code. The efficiency increases with the block size, but as this form of code can correct only a single error, the overall error rate is determined by the probability of two incorrect bits occurring in the same block. This also increases with the size of block so that the maximum permissible frequency of undetected errors will determine the maximum block size and thus the efficiency.

For example, if we have a circuit with a raw error rate of 1 in 10^5 bits and require an uncorrected error rate of around 1 block in 10^8 blocks, we can calculate the probability of two errors in a block (which cannot be corrected) as

4 + 3 block $P = (7 \times 6)/2 \times 10^{-10} = 2.1 \times 10^{-9}$
11 + 4 block $P = (15 \times 14)/2 \times 10^{-10} = 1.05 \times 10^{-8}$
26 + 5 block $P = (31 \times 30)/2 \times 10^{-10} = 4.65 \times 10^{-8}$

The probability of three or more errors is more than three orders of magnitude below this and so can be neglected.

The 11 + 4 block gives an undetected block error rate of 1 bit in 0.952 $\times 10^8$ which just meets the requirement.

This calculation deals only with the particular case of errors which are of short duration and affect only a single bit. At high data transmission speeds errors may last for more than 1 bit time and to obtain a realistic estimate of their effect it is necessary to analyse the nature and duration of faults. One technique used for this is to set up a data generator which produces a sequential series of data items for transmission over the link under test. A similar generator at the receiving end produces the same set of data which is compared with the data coming from the link. All discrepancies are recorded together with a count of the number of blocks received. The appropriate error-correcting procedure can then be designed for each particular link.

9.7 Redundancy schemes

The redundancy schemes used to improve the reliability of data links are classed as temporal since they transmit the same data at different times over the same link. They are effective because the faults in the link are normally impulsive and of short duration and catastrophic failures are much less likely to occur. This form of redundancy costs much less than replication of the complete link (spatial redundancy) and the reduction in capacity caused by the inclusion of extra check bits can generally be accepted. In critical situations the link itself may be duplicated, but the spare circuit is usually employed on a low-priority task which can for a time be suspended when the standby is required.

A comparatively cheap form of redundancy is often used on radio links which suffer from fading due to interference between several signals which arrive over different paths with varying phase shifts. Measurements show that aerials spaced some five to ten wavelengths apart will experience uncorrelated fading. Thus a simple arrangement which continuously selects the largest signal from three aerials suitably spaced will give a much more consistent signal than that obtained from any one aerial. This requires three receiving aerials, three receivers and a little extra electronics for the switching and selection. The most expensive item in the system, the transmitter, is not replicated.

An expensive alternative, which may give a more reliable link, involves frequency diversity. In this the same data is transmitted on different frequencies so giving very low correlation between fading on the different channels.

Bibliography

Bennet, W. R. and Davey, J. R. (1965) *Data Transmission*, McGraw-Hill, New York

Clark, A. P. (1983) *Principles of Digital Data Transmission*, 2nd edn, Pentech Press, London

Clark, G. C. (1981) *Error-Correcting Coding for Digital Communications*, Plenum Press, London

Davies, D. W. et al. (1980) *Computer Networks and their Protocols*, Wiley, London

Hamming, R. W. (1950) Error detecting and error correcting codes. *Bell System Technical Journal*, 29, 147

Jennings, F. (1986) *Practical Data Communications*, Blackwell, London

Morrison, R. (1986) *Grounding and Shielding Techniques in Instrumentation*, 3rd edn, Wiley, Chichester

Parker, J. (1991) Joining electronics with olives. *IEE Review*, 37, No. 10, 347

Peterson, W. W. and Weldon, E. J. (1972) *Error Correcting Codes*, 2nd edn, MIT Press, Cambridge, MA

Sibley, M. J. N. (1990) *Optical Communication*, Macmillan, London

Wiggert, D. (1988) *Codes for Error Control and Synchronization*, Artech House, Norwood, MA

Wilmshurst, T. H. (1985) *Signal Recovery from Noise in Electronic Instrumentation*, Hilger, Bristol

10 Electronic and avionic systems

10.1 Radio transmitters

In the previous chapter the improvement possible in the reliability of a radio link using diversity reception was mentioned. Where the link is a critical part of a control or instrumentation system it may be necessary to provide some degree of redundancy in the transmitting end of the link. The same procedure as used for any complex system can be adopted, that is the division of the system into smaller assemblies, each of which can be replicated. The exact arrangement depends upon the type of transmitter and whether it can be maintained. Generally duplicate redundancy is adequate and in a maintained system both of the duplicate units may normally be active. When a fault appears the faulty unit is disconnected and a reduction in the quality of service is accepted for the short time during which the repair is undertaken.

For an AM transmitter the master oscillator, RF amplifier, modulator and power supply may be duplicated. To avoid synchronizing problems only one master oscillator would normally be used at a time, the second being switched in when needed. Where a single aerial such as a rhombic is used, the outputs of the two amplifiers can be combined and fed to the aerial. Where an aerial array is used this can be split into two sections, each fed from one amplifier. This gives some redundancy in the aerial system, which may be an advantage if it is sited in an exposed position where inclement weather might cause damage.

The general arrangement is shown in Figure 10.1. To cope with more than one failure, facilities can be included to cross-connect, for example RFA1 to MOD2, MOD1 to A2 or PSU1 to either RFA2 or MOD2.

Another technique which has been used to improve reliability is to duplicate the main transmitter components but operate them so that each chain produces half of the rated output. This will give a substantial degree of derating and so will enhance reliability. When one unit fails the other is switched to full power.

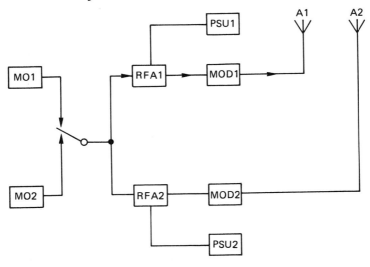

Figure 10.1 Block diagram of duplicate AM radio transmitter

A source of unwanted shut-down in radio transmitters is transients which can produce overvoltages on valves, transmission lines and aerials. These generally cause protective devices such as circuit breakers to trip. The overvoltages can be caused by overmodulation in AM transmitters or lightning discharges to lines or aerials. Although limiters are nearly always fitted to the modulation input chain, short transient overloads do occasionally occur. At maintained stations the first action after such an interruption is to reclose the circuit breaker; in most cases the transient will be over before this and normal operation can be resumed. However, to reduce costs many stations are now unattended, so to prevent lengthy periods of shut-down the circuit breakers are fitted with automatic reclose mechanisms. These will reclose the breaker, say, 5 seconds after it has opened and repeat the process once or twice more. Thus unless the transient fault is repeated several times, the transmitter will only be out of action for a few seconds. This arrangement requires circuit breakers for protection rather than fuses, but gives much better service when short transient overloads are encountered.

Where frequency modulation is used a fault in a duplicate system as described above will not cause a reduction in the received demodulated signal, only a 3 dB reduction in the signal-to-noise ratio. If the link has a reasonable margin of signal strength this should give a very small change in its performance.

One use of radio transmitters which require very high reliability is as radio beacons used by aircraft for navigating along prescribed tracks between airports. These transmitters are the major source of navigational data for aircraft in flight and usually include some form of redundancy.

Another need for highly reliable communication has recently arisen in the provision of UHF and microwave links between offshore oil platforms and the onshore terminals and refineries. These systems are required to handle telemetry data and control signals and are usually equipped with error-correcting facilities and double checking of control commands. Use is also made of satellite links for the longer links (Lakshmanan 1989).

10.2 Satellite links

In non-maintainable systems such as satellite links it is customary to have the standby item unpowered until a fault occurs. It can then be energized and switched into operation, so extending the satellite's life. Typical units duplicated in this way are the low-noise amplifier and the power amplifier. The latter originally used a travelling-wave tube (TWT) which had a limited life and required a high-voltage power supply. Solid-state amplifiers are now available for the 4-6 GHz band which are more reliable, but are also duplicated.

The earliest satellites provided one wideband channel which handled a number of separate signals, including television. Unfortunately the TWT is non-linear, particularly at maximum power output, so causing intermodulation, that is the mixing together of the various signals being handled. To avoid this, current satellites such as INTELSAT V have some 30 different transponders most of which receive 6 GHz signals and transmit back on 4 GHz. Each transponder handles only one signal, so avoiding intermodulation. This kind of problem arises only when analogue signals are transmitted; when using digital coding time division multiplexing can be used so that each input signal has access to the channel in turn for a short interval. As only one signal at a time is handled there is no opportunity for intermodulation and the TWT can be used at maximum power output and so at maximum efficiency. This multiplicity of channels enables a switch matrix to be included in the system between the outputs of the receivers and the inputs of the TWTs. It is thus possible to connect any receiver to any TWT so that by providing a few spare channels, these can act as redundant standby units for the remaining 20 or so working channels. It is also possible to cross-connect these to the 14/11 GHz channels. The switching matrix can be controlled from the earth station, so giving the satellite the ability to continue operation even with several units faulty.

Although early satellites were used mainly to expand the provision of telephone, radio and television links, and thus could not be regarded as of the highest priority, later satellites included some intended solely to aid navigation, controlled by an international organization called INMARSAT.

116 *Reliability in Instrumentation and Control*

These now supply navigational data to some thousands of ships and a failure would put many lives at risk. To reduce the chance of this there are now a number of navigational satellites, and channels are leased from INTELSAT and other satellite operators.

10.3 Aircraft control systems

An important application of complex control systems is in the guidance of aircraft. The earliest systems which took over from the pilot for a short time were used for landing in low visibility. Since a failure could involve major loss of life the Air Registration Board laid down a stringent safety requirement. This was based upon statistics of the accident rate for manually controlled landings in good weather conditions, with a safety factor of 10, leading to a requirement that the probability of a major failure should not exceed 1 in 10^7 landings.

If we assume that the last landing phase in which the aircraft has been committed to the automatic control and the pilot cannot intervene lasts for 30 seconds, the system MTBF needed is 83 500 hours or nearly 10 years. This can be derived from the failure probability which is 10^{-7} for 30 seconds or 1.2×10^{-5} per hour. To demonstrate this MTBF with a confidence level of only 60% requires a test period of 76 500 hours, or nearly 9 years (BS 4200). The extremely high MTBF required for the complete guidance system and the very lengthy testing time make this an impractical scheme. The only way to reduce both of these times is to introduce some degree of redundancy, for example by a triplicated majority voting system. In this scheme the system will operate correctly if any two of the three replicated channels are working. The probability of system failure is the probability that two or three channels will fail. By far the largest of these is the failure of two channels which has a probability of $3p^2$, where p is the probability of one channel failing. Thus $3p^2 = 10^{-7}$, whence $p = 1.83 \times 10^{-4}$ for a 30-second period, or 2.19×10^{-2} per hour. This corresponds to a channel MTBF of only $10^2/2.19 = 45.6$ hours.

This is a much simplified calculation, but it indicates that this form of redundancy can provide an adequate system with a much less onerous target for the channel MTBF and for the time needed to demonstrate it. An alternative scheme which was used in the VC-10 aircraft involved two autopilots, each with a monitoring channel which could detect most errors. The general arrangement of the triplicate scheme is shown in Figure 10.2, and an example of the use of two pairs of processors in the Lockheed L1011-500 airliner is shown in Figure 10.3.

In practice what is classed above as a single channel consists of one complete control system, comprising separate sensing and servo mechanisms to control the elevator, aileron and rudder. Each of these may

Electronic and avionic systems 117

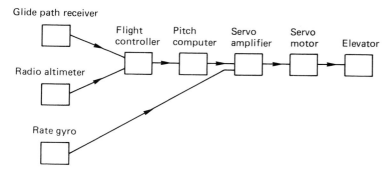

Figure 10.2 *One channel of triplicate elevator control system*

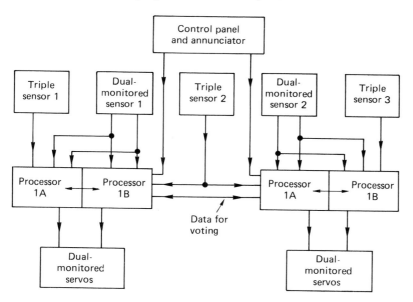

Figure 10.3 *Two pairs of processors used in the Lockheed L1011-500*

comprise some six units, so the complete channel may involve some 18 units. Making the rather simple assumption that they have similar MTBFs, the requirement for each unit becomes about 822 hours. This is not a difficult design parameter for current, mainly solid-state, equipment.

Where the automatic control is required only for landing the rudder control is required only to provide a final 'kick-off', that is a corrective action needed in a cross-wind to align the aircraft along the runway. Prior to this the aircraft must point a little into the wind to maintain its track along the centre line of the runway. The small use of the rudder allows a duplicate form of redundancy to be used without prejudice to the overall system reliability. Where in-flight control is needed the rudder controls will have the same degree of redundancy as the other parts of the guidance system.

There is a further assumption implicit in this calculation which needs examination: the nature of the final voting element. There can be only one of these, and its reliability must be taken into account, since it cannot be replicated and it may determine the complete system reliability. The first systems used three torque-limited motors coupled to the same shaft, which operated the control surface. The shaft was of ample size so that its shearing strength was much greater than the maximum motor torque, and its failure was consequently extremely unlikely.

Later systems replicated the control surfaces, so that the 'voting' operation is the summation of the three forces exerted on the control surfaces by the airflow over them. It is difficult to imagine any way in which this summation could fail so long as the aircraft's speed is maintained: thus this is one case in which the reliability of the voting operation can be taken as 100%.

Later control systems included a requirement that automatic control should be available throughout the flight, with a typical duration of 2 hours. This means a much greater MTBF for each channel, of the order of 11 000 hours.

To diminish the MTBF required for each channel, some use has been made of quadruplex redundancy. In this arrangement several modes of operation are possible; one involves comparing all the channels. So long as they all agree no action is needed. When a disparity occurs the channel whose output disagrees is disabled. If the fault is in the earlier stages of the channel it may be possible to couple the input to the final hydraulic drive to that of another channel, otherwise the control surface can, if possible, be driven to a central position. The three remaining channels can then be configured as a majority voting triplicate system.

The Lockheed L1011-500 airliner is an example of four processors being used in pairs as shown in Figure 10.3. The sensors are connected in either monitored pairs or 2/3 majority logic.

In such a high-reliability application it is essential that great care should be taken to avoid the possibility of common mode faults which could imperil all of the control channels. The most probable of these is a failure of the power supply. To counteract this each engine has an electrical generator, and there is in addition a separate generator independent of the main engines. In addition to electrical supplies a high-pressure oil supply is needed for the actuators which drive the control surfaces. Again each engine has an oil pump and a final standby is provided by an air-turbine-driven pump which can be lowered into the airstream in an emergency.

Some current aircraft which rely heavily on electronic controls rather than mechanical links require very high reliability in the essential flight control systems. For example, the manufacturers of the latest A320 Airbus claim a failure rate of 10^{-9} per hour of flight for the control systems. In the few years it has been in service, however, there were three fatal accidents involving A320s by January 1992. In the crash at Habsheim, France, it could be argued that pilot error contributed to the outcome, but this has not been established for the other two.

In the 1980s NASA adopted a failure rate criterion for aircraft controls of 10^{-9} for a 10-hour civilian flight, for the succeeding decade. This is a target which can only be attained by a high degree of redundancy. In practice on a long-haul flight the pilot can act as a back-up during much of the flight; the only time fully automatic operation is needed is during the landing phase. For example, on transatlantic flights of the Boeing 747 it is customary to switch on only one autopilot during most of the flight; the other two are switched on only for the landing to give full redundancy (Stewart 1984).

No redundancy of controllers will give the required reliability unless the probability of common mode faults such as power failures is very small. To this end it is usual to provide a high degree of redundancy in the supply of electrical power and high-pressure oil for hydraulic servos.

The current plans for the Boeing 777 which should come into service in 1995 provide for a 120 kVA generator on each engine and in the auxiliary power unit. There is also a 30 kVA backup generator on each engine. The captain's primary flight display, navigational display and some engine displays can be supported from only one backup generator. Extra backup is provided for the flaps which are driven electrically if the hydraulic drive fails. Other aircraft use batteries for emergency power supplies.

In addition to the triplicated flight control system modern aircraft such as the Boeing 767 have a duplicate flight management computer which optimizes the total cost, including crew time, maintenance and fuel, a thrust management computer which regulates the automatic throttle and fuel flow, and a dual redundant control display which provides computer-generated flight plans, performance data and advice to the pilot. All of these are connected to a digital bus which also carries the outputs of all subsystems and sensors.

10.4 Railway signalling and control

When railways were first built engine drivers had a large degree of autonomy; there were no timetables and the absence of modern electrical signalling meant that there was no nationwide standard of time. However, as traffic grew it became clear that much more discipline imposed centrally was essential if a safe service was to be offered to the public.

Early accidents caused public alarm and subsequent legislation requiring safety measures such as signal interlocking and improved braking. Braking on early trains was by modern standards extremely poor; brakes were fitted only to the engine and to the guard's van. After an accident in 1889 which killed 78 people when part of a train rolled backwards down a hill and collided with an oncoming train, an Act was passed requiring continuous braking for passenger trains. This involved redundancy in the braking system by fitting brakes to every carriage, so providing a much greater braking force than previously available.

Unfortunately the first brake controls were operated from the engine and when a coupling failed and the train split into two parts the rear portion had no brake power. The need was for some fail-safe mechanism which would automatically apply the brakes if two carriages became uncoupled. After tests it became clear that the Westinghouse vacuum brake was by far the most satisfactory arrangement and this was adopted for all UK trains. The brakes in this scheme are applied under spring loading and a piston fitting inside a cylinder can pull off the brakes when the air is pumped out of the cylinder. The engine provides the necessary vacuum through a pipe which passes along all the carriages.

As soon as two carriages become uncoupled, the pipe is disconnected, the vacuum is broken and the brakes applied automatically under spring pressure. This simple arrangement has proved very reliable over some 100 years of use. Later versions used compressed air for brake actuation to provide greater operating force, but a second sensing pipe was used to ensure that the brakes were applied if two carriages became uncoupled.

As train speeds increased with the development of more powerful engines the reliability of the signalling system became more important. The most likely failure of the mechanical system was a break in the steel cable used to operate the signal semaphore arm. This was originally built so that in the safe condition the cable was not in tension and the arm pointed downwards; to indicate danger the cable was tensioned to bring the arm up to the horizontal position. It was later realized that for a fault involving a break in the wire or the failure of a joint or link (by far the most likely type of failure) this was a 'fail-dangerous' condition since if the signalman set the signal to danger it would revert to clear so allowing a train to pass. The signal aspect was then changed so that the clear position was with the arm pointing upwards and the cable in tension. A break in the cable would then allow the arm to fall to the horizontal position, meaning danger. The break would then cause a 'fail-safe' fault which would stop any train arriving at the signal.

Even greater train speeds caused difficulties in seeing the signals and four-aspect colour signals were introduced. These do not involve any movement of a signal arm and there is no possibility of a fail-safe design based on the operation of a mechanical link. The signals use electric filament lamps and the reliability of the signalling depends mainly upon the probability that the lamps will not fail. Even when derated with the penalty of reduced light output the life of a filament lamp is only some 4000 hours so that, used alone, it would be unacceptable. To give a much better life redundancy is used in the form of a twin-filament lamp. The current taken by the main filament is monitored and when it fails the second filament is automatically switched on and a relay gives a fault indication in the signal box so that the lamp can be changed as soon as possible. If the second filament fails, the nearby signals are automatically set to danger. Thus as well as using redundancy the final backup is the fail-safe action of stopping all oncoming trains.

Despite the redundancy and the fail-safe features of this form of signalling its effectiveness depends on the driver's ability to interpret correctly the information given by the signals. Thus any further progress requires some backup for the driver. This was first introduced by the GWR in the 1930s by using a movable ramp between the rails which gave an audible warning in the cab when the train passed a distant signal set at danger. This had to be cancelled manually by the driver within a short period, otherwise the brakes would be applied automatically.

A similar scheme was later adopted nationally, but with induction loop signalling between the track and the train. The London Underground system has a much greater traffic density than the mainline railways, trains following one another at intervals of 2 minutes or less. A backup system which relies on the driver to apply the brakes may involve unnecessary delay, so on the latest extension to the Underground, the Victoria line, the track signals control both the train speed and the brakes.

As it is likely that further developments in automatic train control will involve microprocessors, some development work was initiated by British Rail into high-reliability systems. The scheme they selected used twin processors which had identical inputs from axle-driven tachometers and voltages induced by track-mounted cables and magnets (Cribbens 1976). The processors are cross-connected and have identical inputs. Periodically each one compares its output states with those input from the other processor and any disagreement is interpreted as a fault and the check result line is energized. This shuts down the system permanently by blowing a fuse which also removes the supply from any displays. This scheme can easily be extended to a triplicate scheme in which any disagreement will shut down only the faulty processor. The remaining two can continue operation so long as their outputs agree.

10.5 Robotic systems

The increasing use of automation in manufacturing processes has led to the use of a large number of industrial robots. By 1985 some 100 000 were in service, mainly in Europe, the USA and Japan. In the interest of safety they should be segregated from human operators, but in many cases this is not completely possible since they have to be loaded with workpieces and the finished product removed. A further problem arises with robots which have to be 'taught' what to do. These are used in work such as paint spraying in which it is impossible to calculate the path along which the robot hand should move. The robot is provided with internal storage capacity and the hand is moved under manual control to simulate the action of a human operator by using a small hand-held control box. The trajectory of the hand is then stored within the robot and can be repeated automatically as often as required.

Clearly this means that someone must be inside the robot enclosure when the power is applied and it is essential that this person's safety is ensured at all times. Also it may not be possible to maintain and test the robot properly without touching it.

Anyone approaching the robot will have no knowledge of the path the arm is programmed to follow, and there is no external indication of this. It is thus liable to make sudden and unexpected moves. Furthermore, if the power suddenly fails it may drop its load.

All of these factors make the reliability of the robot control system (generally based on a small computer) of extreme importance. Redundancy can be used with either two or three channels. With two channels the two output signals can be continuously compared; any disparity immediately stops the robot's motion. Three-channel redundancy can be arranged as a majority voting system. Other safety features which can be included are:

1. In the 'teaching' mode the linear speed of the arm can be held down to, say, 20 cm/s to avoid danger to the operator.
2. An area extending at least a metre beyond the robot can be fenced off to prevent access.
3. The robot arm and other moving parts can be equipped with touch-sensitive pads which sense when anything is touched and so halt the robot.
4. Emergency buttons can be fitted which stop the pump and dump the high-pressure hydraulic supply.
5. Built-in test equipment can be fitted which checks most of the robot control system each time power is switched on. Any fault will prevent the robot's motion.

Bibliography

Andeen, G. B. (ed.) (1988) *Robot Design Handbook*, McGraw-Hill, New York

Bernhard, R. (1980) All digital jets on the horizon. *IEEE Spectrum*, October

Bonney, M. C. and Yong, Y. F. (eds) (1985) *Robot Safety*, IFS, Bedford
Brady, M. (ed.) (1982) *Robot Motion, Planning and Control*, MIT Press, London
Chetty, P. R. K. (1988) *Satellite Technology and its Applications*, Tab Books, Blue Ridge Summit, PA
Cribbens, A. H. *et al.* (1976) The microprocessor as a railway control system component. *Microprocessors* **1**, no, 1, 41
Dalgleish, D. J. (1989) *An Introduction to Satellite Communications*, Peter Peregrinus, London
Feltham, R. G. (1973) *Aircraft Accident Data Recording Systems*, HMSO, London
Ha, T. T. (1990) *Digital Satellite Communications*, 2nd edn, McGraw-Hill, New York
Hunt, V. D. (1983) *Industrial Robotics Handbook*, Industrial Press, New York
Lakshmanan, K. (1989) Plumbing the depths. Telemetry and telecontrol in offshore oil installations, *IEE Review*, Dec.
McCloy, D. and Harris, M. (1986) *Robotics: An Introduction*, Open University Press, Milton Keynes
Mancing, V. J. *et al.* (1986) Reliability considerations for communications satellites. *Proceedings of the Association of Reliability and Maintainability Symposium*
Martin, D. J. (1982) Dissimilar software in high integrity flight controls. *AGARD Symposium on Software for Avionics*, The Hague
Punches, K. (1983) Airplane reliability in a nutshell. *IEEE Transactions on Reliability*, **R-32**, 130
Roberts, R. S. (ed.) (1985) *TV Engineering: broadcast, cable and satellite*, Pentech Press, London
Sonnenberg, G. J. (1978) *Radar and Electronic Navigation*, Butterworths, London
Stewart, S. (1984) *Flying the Big Jets*, Airlife Publishing, Shrewsbury
Tetley, L. (1986) *Electronic Aids to Navigation*, Edward Arnold, London
Wensley, J. H. (1979) SIFT – Design and analysis of a fault-tolerant computer for aircraft control. *Proceedings of the IEEE*, **66**, 1221
Wertz, J. R. (ed.) (1980) *Spacecraft Altitude Determination and Control*, Reidel, London

11 Nuclear reactor control systems

11.1 Requirements for reactor control

In the design of many control systems the importance of reliability depends upon the consequences of a failure. The cost of a failure in the control circuits of a domestic appliance such as a washing machine is largely restricted to the cost of repair; this will fall on the manufacturer if it occurs within the guarantee period, after this on the owner. The manufacturer is interested in making a product with sufficient reliability to ensure a very low level of returns under guarantee and to keep up with the competition.

At the other end of the spectrum the cost of a failure of the control or safety systems of a nuclear reactor is extremely large but not easy to quantify. However, the consequences of the Chernobyl accident have been very great in the destruction of equipment, the sterilization of land and illness and loss of life. In 1957 an American report (WASH-740) predicted that the cost to property alone in a worst-case accident could be \$7 billion; this was updated in 1964-5 to a figure of \$40.5 billion (Curtis and Hogan 1980) and would be much larger at today's prices. The accident at Three Mile Island in the United States in 1979 caused a loss of coolant and a partial melt-down of the core. Although there were no casualties and no one off the site received excessive radiation the cost to the owners was considerable. The TMI-2 reactor was disabled and the Nuclear Regulatory Commission would not allow the twin TMI-1 reactor to restart until the accident had been investigated.

In consequence of this extremely high cost the reliability specified for reactor safety systems is also very high; the US Nuclear Regulatory Commission in 1986 made a policy statement about future advanced reactors, which were expected to show a considerably better safety record than existing reactors and have a probability of less than 10^{-6} per year (1.14 $\times 10^{-10}$ per hour) for excessive radionuclide releases following a core melt-down (Okkrent 1989).

France and Switzerland have safety targets similar to that of the USA and a similar target has been set in the UK for the core-melt frequency of the

Sizewell B station. Sweden has a similar target for core melt, but the probability of severe radioactive contamination of large land areas is required to be even lower. The Swedish firm ASEA-Atom has proposed an inherently safe reactor (PIUS) which can be abandoned at any time by its operators and will then automatically shut itself down and stay safe for about a week. The core and the primary coolant system is surrounded by a large pool of borated water and any system upset will cause the borated water to flood into the reactor coolant system. As boron is a strong neutron absorber, this will shut the reactor down and it is also designed to establish natural convection cooling (Forsberg 1986). Another automatic shut-down mechanism has been proposed for a power reactor inherently safe module (PRISM) in which any upset which causes a considerable increase in the coolant temperature would automatically cause safety rods to fall into the reactor core. The unit consists of nine reactor modules feeding three turbine generators. Each module has six neutron-absorbing rods for power control and three articulated rods suspended by Curie point magnets. These are for emergency shut-down. When the magnets reach their Curie temperature, they lose their magnetism and the rods fall into the core (Van Tuyle et al. 1990).

11.2 Principles of reactor control

The reliability required for a nuclear power station control system is almost certainly greater than that required in other systems, as can be seen from the specified failure rates quoted in the previous section. This reliability involves a high degree of redundancy in the control and safety systems and calls for particular features in the design of the reactor itself.

It is interesting to recall that the first reactor which demonstrated self-sustaining nuclear fission, the Enrico Fermi pile built at the University of Chicago, had primitive but redundant safety measures. A large neutron-absorbing boron rod was suspended by a clothes line above the pile and a graduate student was given an axe and told to cut the line, so releasing the rod, in the event of an accident. Also another group of students were stationed above the pile and given bottles of gadolinium (another neutron absorber). If a problem arose, they were told to throw the bottles and run away.

Two major concerns in the design of the reactor are to prevent the escape of radioactive material into the surroundings and to prevent a core melt-down. As far as possible the reactor is designed to provide at least two methods of coping with any failure. To prevent the escape of radioactive material the fuel is contained in closed cans and the entire reactor core is enclosed in a large enclosure – concrete in the early gas-cooled reactors and steel in light-water reactors.

Also, to prevent the failure of a coolant pump causing overheating of the core several pumps are used and the reactor can survive the failure of any one. For example, the early Magnox stations had six gas circulators and the later AGRs such as Hinkley have eight. Also the loss of the power supply for the pumps will automatically shut down the reactor.

The choice and location of the sensors used to monitor the state of the reactor is decided largely by the principle that any conceivable failure of the reactor system should be detected by at least two types of instrument. Thus excess activity within the reactor will be shown by an increase in fuel can temperature and also by excess neutron flux. A rupture in the coolant circulation pipes of a gas-cooled reactor would be revealed by a large rate of change of gas pressure (dp/dt) and by changes in the channel outlet gas temperature.

One design factor which will make the control task much easier is to choose a combination of fuel, coolant and moderator which has a degree of inherent stability. This is usually a matter of minimizing fluctuations of fuel temperature, by selecting a combination which has an overall negative temperature coefficient of reactivity. This means that if a disturbance causes an increase in neutron flux and thus in core temperature, the result is a decrease in reactivity which will limit the disturbance. This is a form of negative feedback.

The early Magnox gas-cooled reactors had a variety of sensors for detecting faults, mostly arranged in 2/3 majority voting logic, as shown in the following table.

Quantity sensed	No. of channels	Logic scheme
Fuel can temperature	9	$(2/3)^2$
Rate of change of pressure	6	$2 \times 2/3$
Channel outlet gas temperature	36	$12 \times 2/3$
High-power excess flux	9	$3 \times 2/3$
Low log excess flux	3	2/3
High log doubling time	3	2/3
Low log doubling time	3	2/3
Loss of blower supply	6	3/6

The later AGR reactors had 27 different parameters measured, all arranged in 2/4 logic, and proposed PWRs will have a few more parameters in the primary system with nine additional parameters being measured in the secondary protection system. All of these are connected in 2/4 logic (Jervis 1984).

128 *Reliability in Instrumentation and Control*

The main limitation on the reactor output is the maximum permitted fuel can temperature. In order to regulate this the neutron-absorbing control rods are adjusted and this requires a large number of thermocouples which measure can temperature. Because these deteriorate with high temperature and irradiation they are usually replicated. In some reactors they are fitted in groups of 16. The signal amplifier scans them and selects the element which shows the highest temperature to indicate the spot temperature.

To ensure the necessary high reliability in the control function, and particularly in the provision for automatically shutting the reactor down under fault conditions, a high degree of redundancy is used, generally triplicated with majority voting. This is used either singly or in two stages, starting with nine separate inputs, as shown in Figure 11.1. For higher reliability 2/4 schemes have also been used, for example in the Heysham 2 AGR.

For a 2/3 majority voting scheme the output will be incorrect if either two or three of the inputs are incorrect. If the probability of a false input is p, the probability of a false output is

$$p_1 = 3p^2 + p^3 \tag{11.1}$$

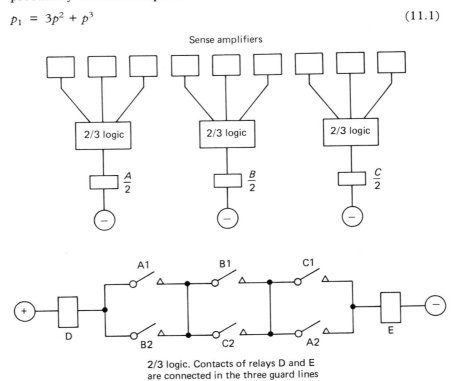

Figure 11.1 Double 2/3 logic

Since p is normally small the second term is much smaller than the first and it can be neglected in view of the uncertainty associated with the value of p. For a two-stage version, often called $(2/3)^2$ we can use Equation (11.1) again to give

$$P = 3p_1^2 = 3(3p^2)^2 = 27p^4 \tag{11.2}$$

This result ignores the failure probability of the voting circuit which should be added if it is significant.

The use of 2/4 logic improves the reliability of the redundant system with regard to dangerous failures, but is rather worse than 2/3 logic with regard to safe failures. Thus in order to prevent the correct response to a reactor fault three or four channels of a group must fail. If the probability of one channel failing is p, the probability of a dangerous failure is then $P_D = 4p^3 + p^4$. Generally the second term is much less than the first, so that the value can be taken as $4p^3$, compared with $3p^2$ for 2/3 logic.

However, if only two of the channels give an incorrect 'trip' indication, the reactor will be shut down, giving a safe failure. The probability of this is $P_S = 6p_1^2$, where p_1 is the probability of a channel failing safely.

The failure rates of solid-state equipment in nuclear power stations have been found to lie in the range of 0.01 to 0.1 failures per year (Eames 1978). Taking an average value of 0.05 gives a failure probability of 0.0125 for a 3-month interval. This is of interest as the licence for the early reactors required all monitoring equipment to be tested regularly every 3 months.

This figure is for all failures; the critical ones are the dangerous failures which on average form about one-third of the total. Thus the dangerous failure rate is 0.0042 over 3 months. In a $(2/3)^2$ arrangement this gives a failure probability of $27(0.042)^4 = 8.40 \times 10^{-9}$.

A dangerous failure can occur at any time during the maintenance cycle and it will persist until the next test/calibrate/repair action. Thus the fault can remain for a period varying from zero to the maintenance interval, in our case 3 months. The average time is thus 1½ months.

In order to assess the consequent likelihood of a severe accident we need also to know the chance of a reactor having an upset which would call for the action of the safety system. This is not an easy figure to discover, but some guidance is available from American sources; these show (Phung 1985) that 52 significant events occurred between 1969 and 1979 during about 400 reactor years of operation. This is a frequency of about 1 in 8 reactor years, or 0.0156 in 1½ months.

Some later information published in 1989 deals with seven events during some 940 reactor years of BWR operation; this corresponds to a frequency of 1 in 134 reactor years (Burns 1989), but deals with only one type of reactor. There is a considerable discrepancy between this and the previous figure, but at least it suggests that 1 event in 8 reactor years is a conservative assumption.

The probability of a simultaneous reactor upset and a dangerous failure of the monitoring equipment is then

$8.4 \times 10^{-9} \times 0.0156 = 1.31 \times 10^{-10}$

This calculation relates to the interval between maintenance actions – 3 months. We assume that after service the equipment is as good as new, so that each 3-monthly period has the same probability of failure. Thus the probability of a reactor upset which will find the monitoring equipment in a fail-dangerous state is

$4 \times 1.31 \times 10^{-10} = 5.24 \times 10^{-10}$ per year

This result relates to only one group of sensors, but even with 100 groups the probability is well below the target figure of 10^{-6} per year and allows for some common mode failures.

Using 2/4 logic with the same fail-dangerous probability per channel per 3 months of $p = 0.0042$ gives a group failure rate of

$p_1 = 4p^3 = 4 \times (0.0042)^3 = 2.96 \times 10^{-7}$

Combining this with the probability of a reactor upset of 0.0156 per 3 months gives an overall probability of a dangerous fault as 4.62×10^{-9} over 3 months, or for a year the figure of 1.85×10^{-8}. This configuration of double 2/4 logic was used for the later AGR stations which included solid-state monitoring units and these are likely to have a somewhat lower failure rate than is assumed in the above calculation. In the double 2/4 or $(2/4)^2$ configuration the first stage of logic gives $p_1 = 4p^3$. The second stage gives $P = 4p_1^3 = 256p^9$. This configuration is used in the Heysham 2 power station.

The periodic testing of the reactor protective system requires skilled technicians and is a rather lengthy process. Experience with French reactors shows that this requires two technicians and takes some 10 hours to complete. The same operation prior to restarting the reactor requires 2 days and involves a significant delay in restarting the reactor after a shut-down (Raimondo 1990). In order to improve this activity attention is now being devoted to the automatic testing of all sensing amplifiers and signal conditioning equipment. In most nuclear power stations the control and safety shut-down systems are separate. The safety circuits and their logic are normally hardwired, but increasingly the control function is handed over to a computer. This basically involves regulating the reactor output so that the heat output and the electrical power generated match the load demanded by the transmission system which the station feeds. This is a tedious and difficult task to perform manually and a computer can ease the burden on the operator. The major problem is the need to control various sectors of the reactor to ensure a fairly even temperature distribution throughout the core. Early stations used a central computer with standby, but later AGR stations such as Heysham 2 used 11 distributed microprocessors with direct digital control to replace the former analogue control loops. The advantage of the digital technique is that more complex control algorithms can be used and it easier to modify and optimize the various control loops. The availability of computing power enables some of the safety monitoring to be undertaken by the computer, the rest being performed by hardwired logic. This gives a useful diversity of techniques for the protection circuits.

11.3 Types of failure

The distinction between safe and dangerous failures is of critical concern in nuclear power stations; a safe failure will shut the reactor down without cause. This involves a considerable loss of revenue as most nuclear stations are used to supply the base load and are run as nearly as possible at full power continuously. A further problem arises from the build up of neutron-absorbing products such as xenon after a reactor has been shut down. This element builds up for about 10 hours and then decays, effectively 'poisoning' the reactor so that unless it is restarted soon after shut-down the operators must wait for perhaps 20 hours before there is sufficient reactivity to permit sustained fission.

The accident at Chernobyl has shown the world the enormous cost of a nuclear melt-down which emitted a large volume of radioactive products. Thus there is a great impetus to make all monitoring equipment very reliable, and to ensure that as many of the failures as possible are safe failures. Some steps can be taken in the design of the equipment to this end, but it is impossible to eliminate all dangerous faults.

If we use directly coupled amplifiers, either with discrete components or as IC operational amplifiers in general, faults will result in the output either falling to a voltage near to the lowest power supply or rising towards the most positive supply. One of these will represent a safe failure and the other a dangerous failure, and they will usually occur in equal proportions. One method of reducing the proportion of undetected dangerous failures is to convert the incoming signal to a square wave which is subsequently amplified by an AC amplifier and rectified. Almost all faults will result in the disappearance of the AC signal and this can be monitored and cause an alarm.

The early Magnox stations used relays for the logic of the safety system; they are fast enough and regarded as very reliable. To reduce the chance of dangerous failures they were connected so that the most reliable operation (opening a normally closed contact) corresponds to the action needed to cope with a reactor fault. The contacts of the various monitoring groups are connected in series with two contactors and the power supply, and all contacts are normally closed. The series circuit is triplicated. Two guard

132 *Reliability in Instrumentation and Control*

relays are used in each guard line, one next to each pole of the power supply to ensure that an earth fault cannot cause a dangerous failure. The contacts of the guard relays are connected in a 2/3 circuit to power contactors which are also normally operated and connect a power supply to the electromagnets which hold up the reactor control rods. The scheme is shown in Figure 11.2 (Weaving and Sherlock 1963). When a fault occurs in the reactor the current in all relay and contactor coils is cut off and their contacts open. The final result is that the control rods are released from their supports and fall into the reactor core, so reducing the neutron flux and the generation of heat.

Although the telephone type relays originally proposed for these circuits have a long history of reliable operation, some design engineers thought that this could be improved. Their object was to remove a possible source of failure – pickup of dust and contamination by the contacts – by sealing the relay in an airtight enclosure. This should in principle afford some improvement, but in fact it produced a number of dangerous failures which were detected during routine tests in which a relay occasionally failed to open its contact when de-energized. In order to investigate the fault the relay was removed and the enclosure cut away. Unfortunately the handling disturbed the relay and cleared the fault and it was necessary to take an X-ray photograph of the relay *in situ* to discover that the relay contacts seemed to be stuck together. This was traced to the effect of the varnish used to impregnate the relay coil. Although the coil is baked to dry the varnish before the relay is assembled it is very difficult to remove all traces of solvent. As the relay coil is normally energized when it is in use the heat evolved would evaporate any remaining traces of solvent and these were found condensed on the contacts. The problem was solved by sealing the contact stack separately so that no contamination from the coil could reach them.

11.4 Common mode faults

The need for extremely high reliability in nuclear safety systems requires great care to avoid any common mode faults. The first consideration is the integrity of the power supplies; the arrangement of the relay logic is basically fail safe in that the removal of the power supply would shut the reactor down. However, this will involve considerable loss of income since nuclear plant is generally used to supply base load and is thus delivering near to full output for 24 hours a day. It is thus worth providing typically two different rectified AC supplies, a standby battery and a separate generator with automatic switching to ensure a very low probability of loss of power. Another possibility is to supply the equipment from float-charged batteries which will ensure operation for some hours if there is a mains

Nuclear reactor control systems 133

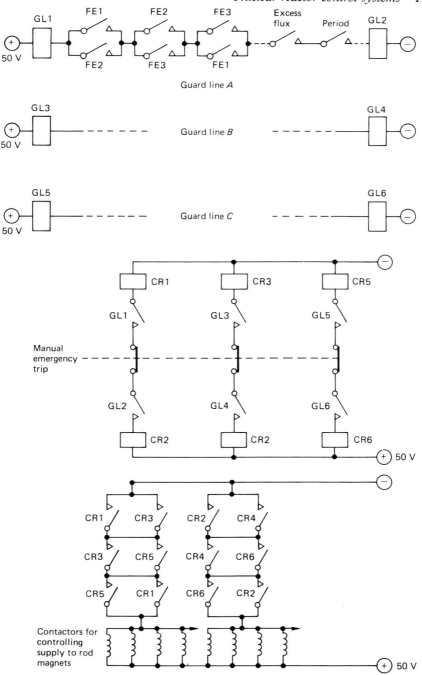

Figure 11.2 *Redundant logic controlling safety shut-down rods*

failure. Usually a standby engine-driven generator is started a few minutes after a mains failure has occurred to continue float-charging. There may also be a second generator or an alternative mains supply.

Another possible common mode failure was revealed by the Three Mile Island accident; the power and control cables were all taken through the same tunnel between the plant and the control room. Consequently when a fire occurred initially in the power cables this very soon damaged the control cables, so preventing important details of the plant's condition from reaching the operators. Some of the safety circuits were also damaged. It is essential that the redundancy in the instrumentation should not be nullified by bunching together all power and signal cables in the same duct or trunking. Ideally each of the three or four channels involved in every measurement should be physically separated from the others, with an appropriate fireproof barrier. A worst-case possibility is a major damage to the main control room; to cope with this a separate emergency standby facility may be provided some distance away with minimum instrumentation and sufficient controls to monitor the reactor state and shut it down safely. For example, at the Heysham 2 AGR station the emergency centre is some 100 m away from the main control room.

11.5 Reactor protection logic

The early Magnox stations used standard telephone-type relays for switching logic; their characteristics has been established over some decades of use and their reliability was considered to be adequate. However, as the monitoring equipment became more reliable, mainly through the change from valves to transistors, a comparable increase in the reliability of the switching logic was sought. One possibility was to move to some static apparatus rather than electromechanical devices such as relays. Early transistors had poor reliability and magnetic devices offered much higher reliability, consisting only of magnetic cores and copper windings.

The design principles adopted were as follows:

1 Each parameter checked by the safety system should be measured by three separate channels whose outputs are combined on a majority voting basis. If only one channel trips an alarm should be given.
2 No single fault or credible combination of two equipment faults should prevent the reactor being shut down when a demand occurs.
3 No single fault in the safety circuits should cause a reactor trip.
4 If three guard lines are used they should all be opened by a reactor fault condition, if all safety equipment is working correctly.
5 Once a guard line has tripped it should remain tripped until restored manually.

6 The guard lines should be segregated from one another and only one should be accessible at a time.
7 A shorting plug is provided which can be inserted to maintain the continuity of the guard line when any piece of equipment is removed for maintenance. Only one item should be able to be removed at a time.
8 The control rods should be operated by two distinct methods. This is usually ensured by providing two groups of rods, controlled separately but by similar means.
9 To allow for overriding operator action a manual trip should be provided which is as close as possible to the final control rod holding circuits.

The component selected for the logic function was a multi-aperture ferrite device known as a 'Laddic' (Weaving and Sherlock 1963). This had the geometry of a ladder with several rungs, with a number of different windings. After experiments a seven-aperture device was adopted. This is energized by two interleaved pulse trains at a frequency of 1 kHz called 'set' and 'reset'. The pulses required for each guard line are about 1 A peak with a duration of 10 µs. An output is obtained only when the DC 'hold' signals are present, so giving a three-input AND logic function. The hold currents are obtained directly from the various monitoring units, and are cut off when any parameter reaches its threshold value. By splitting the hold windings in two, either of them can be energized to provide an output, so giving an OR logic operation. With suitable connections, the AND and OR functions can be combined to give an output $X = (A + B).(B + C).(C + A)$. This is logically equivalent to $X = A.B + B.C + C.A$ which is a 'two out of three' majority vote, the function required for a triplicated guard line. The arrangement of core and windings is shown in Figure 11.3. The output signal from the Laddic is a pulse of about 100 mA peak and 2 µs long. It requires only a single-transistor amplifier to couple this to the next Laddic when a chain of them is used to form a guard line.

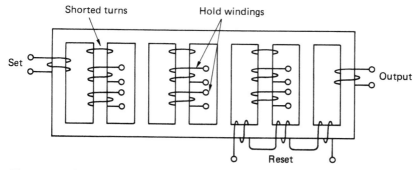

Figure 11.3 *Core and windings of Laddic*

As the output is a pulse train, a pulse-to-DC converter is used to turn this into a continuous signal which can hold up the safety rods. The converter does not reset automatically, so that once its output has disappeared it will not return until a manual reset is operated.

One way of guarding against common mode failures is to use more than one set of safety equipment, with different technologies and physically separated. This concept led to the development of pulse-coded logic for reactor safety circuits (Fast 1987) which initially used standard integrated circuits and was designed for 2/3 voting. Three guard lines are used in a 2/3 configuration so that if any two or all three guard lines trip all six shut-down rods fall into the core as the supply to the magnets which holds them up is cut off. Each guard line has its own pulse generator which generates a train of pulses which examine the state of each trip instrument. The three pulse generators are synchronized and the pulse train from each line is communicated to the others via optical fibres.

In one version of the scheme the pulse trains consist of 10 bits at 7.8 kbits/s and a 'stuck at 1' or a 'stuck at 0' fault in any instrument will produce a unique code which can be used to identify the faulty unit. A 14 parameter version of the scheme was attached to the Oldbury power station protection system between 1978 and 1982 and then transferred to the DIDO materials testing reactor, acting in a passive role. It has now been expanded for active operation, and is considered as adequate for use in a commercial power reactor.

Bibliography

Asahi, Y. *et al.* (1990) Conceptual design of the integrated reactor with inherent safety (IRIS). *Nuclear Technology*, **91**, July, 29

British Nuclear Energy Society (1987) *Science and Technology of Fast Reactor Safety, Proceedings of Conference, 1986*, Vols 1 and 2, BNE Soc, London

Burns, E. T. (1989) Reassessment of the BWR scram failure problem. *Transactions of the American Nuclear Society*, **59**, 195

Corcoran, W. R. *et al.* (1981) The critical safety functions and plant operation. *Nuclear Technology*, **55**, December

Curtis, R. and Hogan, E. (1980) *Nuclear Lessons*, Turnstone Press, Wellingborough

Eames, A. R. (1978) Prediction methods for equipment reliability evaluation. *Radio & Electronic Engineering*, **48**, 333

Farmer, F. R., (ed.) (1977) *Nuclear Reactor Safety*, Academic Press, London

Fast, B. (1987) A reactor protection system using pulse coded logic. *Science and Technology of Fast Reactor Safety*, Vol. 2, BNES, London

Forsberg, C. W. (1986) A process inherent ultimate safety boiling water reactor. *Nuclear Technology*, **72**, 121

Green, A. E. (ed.) (1982) *High Risk Safety Technology*, Wiley, Chichester

Haas, P. M. and Bott, J. E. (1980) Criteria for safety-related nuclear plant operator actions. *6th Advances in Reliability Technology Symposium, 1980*, UKAEA, Warrington

Jervis, M. W. (1984) Control and instrumentation of large nuclear power stations. *IEE Proceedings*, **131**, Pt A, 481

Keats, A. B. (1980) Fail-safe design criteria for computer-based reactor protective systems. *Nuclear Energy*, **19**, No. 6

Moss, T. H. and Sills, D. L. (eds) (1981) The Three Mile Island nuclear accident: Lessons and implications. *Annals of the New York Academy of Sciences*, **365**

Okkrent, D. (1989) A look at safety goals and safety design trends for advanced light water power reactors. *Nuclear Technology*, **88**, 166

Patterson, D. (1968) Application of a computerised alarm analysis system to a nuclear power station. *Proceedings of the IEE*, **115**, 1858

Phung, D. L. (1985) Light water reactor safety before and after the Three Mile Island accident. *Nuclear Science and Engineering*, **90**, 509

Raimondo, E. (1990) Automatic reactor protection testing saves time and avoids errors. *Nuclear Engineering International*, April, p. 21

Rubinstein, E. (1979) Special Issue: Three Mile Island and the future of nuclear power. *IEEE Spectrum*, **16**, 30

Todd, S. N. (1981) The use of microprocessors in gas-cooled nuclear reactors. *Electronics & Power*, **27**, 323

Tylee, J. L. (1983) On-line failure detection in nuclear power plant instrumentation. *IEEE Transactions on Automatic Control*, **AC-28**, 406

Uhrig, R. E. (1989) Automation and control of next-generation nuclear power plant. *Nuclear Technology*, **99**, 157

Van Tuyle, G. J. *et al.* (1990) Examining the inherent safety of PRISM, SAFRS and the MHTGR. *Nuclear Technology*, **91**, No. 2, 185

Weaving, A. H. and Sherlock, J. (1963) Magnetic logic applied to reactor safety circuits. *Journal of the British Nuclear Energy Society*, **2**, 74

Welbourne, D. (1968) Alarm and display at Wylfa nuclear power station. *Proceedings of the IEE*, **115**, 1726

12 Process and plant control

12.1 Additional hazards in chemical plant

Although the equipment mentioned in previous chapters is often required to operate in hostile environments involving salt spray, dust, humidity, etc., the atmosphere which envelops it is generally inert. In contrast, many chemical plants handle corrosive, toxic, flammable or explosive substances, and any design for reliability must tackle these extra hazards. This affects two system areas; firstly the major plant activity which usually involves transporting liquid or gaseous materials around the plant while it undergoes various processes. The probability of various modes of failure must be examined and appropriate methods devised to counteract them. This often involves providing alternative ways of performing various operations.

At the same time most of the control and measurement information is handled as electrical signals and output devices may include solenoid- or motor-operated valves and electrically driven pumps. Any switch, relay or contactor which involves moving contacts and handles an appreciable current can cause a spark when interrupting current. The temperature of the spark can reach thousands of degrees centigrade, sufficient to ignite any flammable gas or vapour. Thus any such equipment used in a location where flammable gas or vapour may be present must be surrounded by a flameproof enclosure (Bennett 1988, Sansom and Franklin 1988). The enclosure must have suitable access covers which can be removed for inspection, maintenance or the connection of power supply cables. The covers must have sufficiently wide flange couplings, and small enough airgaps to prevent a flame propagating from the inside of the enclosure to the outside and so causing a fire or an explosion. In addition to this requirement for normal working conditions it is necessary to ensure that no accident can occur when the covers are removed. The main possibilities arise from 'hot spots', that is parts of the equipment which are hot enough to ignite a flammable gas or vapour, and the discharge time of capacitors. There must be enough delay between switching off the power and allowing access to capacitors to ensure that they have no residual charge which could cause a spark. BS 5501 (EN 50014) specifies the maximum charge on the capacitors which is permitted when the enclosure is opened. If the charge has not decayed to a safe value when the case is opened normally a label is required showing the delay needed after disconnecting supplies before the case should be opened. A similar delay may be needed if any component has an excessive surface temperature, to allow it to cool.

BS 5000 and BS 5501 specify various ways in which electrical apparatus can be made safe for use in explosive atmospheres. These include:

- Type 'd' Flameproof enclosures. These can withstand internal explosion of an explosive mixture without igniting an explosive atmosphere surrounding the enclosure.
- Type 'e' Increased security against possibility of excessive temperature and occurrence of arcs and sparks.
- Type 'i' Intrinsic safety. This specifies electrical systems in which the circuits are incapable of causing ignition of the surrounding atmosphere. The maximum nominal system voltage is restricted to 24 V (limit of 34 V) and a normal current of 50 mA, short-circuit value 100 mA.
- Type 'm' Encapsulated to prevent contact with atmosphere.
- Type 'p' Pressurized enclosure. In this a protective gas (usually air) is maintained at a pressure greater than that of the surrounding atmosphere. The enclosure must withstand a pressure of 1.5 times the internal pressure.
- Type 'o' Oil immersed. Here the oil will prevent any arcs or sparks igniting an external flammable gas or vapour.
- Type 'q' Powder filling. This specifies apparatus using voltages of 6.6 kV or less with no moving parts in contact with the filling. The preferred filling is quartz granules; no organic material is permitted.

The hazard posed by liquids is mainly dependent upon their flash point; those with flash points below 66°C are classed as flammable and those with flash points below 32°C as highly flammable. Many industrial gases and vapours will ignite with a concentration of only 1% by volume in air, and some mixtures are flammable with a wide range of concentrations. For example, any mixture of air and hydrogen with between 4% and 74% of hydrogen by volume is flammable. For a mixture of acetylene and air the limits are 2.5% and 80%. The limits for an explosive mixture are generally somewhat narrower than for flammability.

12.2 Hazardous areas

An important factor in designing equipment for use in areas which may contain flammable gases or vapours is the distance from the point of release within which a dangerous concentration may exist. Some degree of classification is given in BS 5345: Part 2 (IEC 79-10). The grade of release is designated as continuous where it is expected to occur for fairly long periods, as primary where it is expected to occur periodically or occasionally during normal working, and as secondary where it is not expected to occur in normal operation, and if it does so only infrequently and for short periods.

Areas of continuous release are normally graded as Zone 0, areas of primary release as Zone 1 and areas of secondary release as Zone 2.

BS 5345 also classifies the ventilation, an important factor in deciding the extent of the hazardous areas, according to its effectiveness in clearing away dangerous gases or vapours and the proportion of working time during which it is in operation (Townsend 1988).

An area classification code applied to petroleum has been proposed by the Institute of Petroleum (Jones 1988). This puts liquefied petroleum gas in category 0 and classifies other petroleum products according to their flash points (FP):

- Class I: liquids with FP below 21°C.
- Class II(1): liquids with FP from 21°C to 55°C, handled below their FP.
- Class II(2): liquids with FP from 21°C to 55°C, handled at or above their FP.
- Class III (1): liquids with FP above 55°C to 100°C, handled below their FP.
- Class III (2): liquids with FP above 55°C to 100°C, handled at or above their FP.
- Unclassified: liquids with FP above 100°C.
- A further area classification specifically for all installations handling natural gas at all pressures is based upon the British Gas Engineering Standard BG/PS/SHA1 (Gregory and Brown 1988).

A factor which greatly affects the hazard posed by gases or vapours is their density relative to air. If they are heavier than air they will tend to collect near the ground and are much more likely to ignite than lighter gases such as hydrogen which will rise and disperse readily, particularly out of doors.

12.3 Risks to life

Inadequate reliability of plant and equipment can involve high expenditure in repairing damage caused by failure and loss of income while the plant is inoperative. Increased expenditure on measures to improve reliability can be justified if their cost is less than the reduction they are likely to bring in the cost of failure. This calculation is not difficult although the data required may be known only imprecisely. For example, if the probability of a dangerous occurrence is estimated as once in 400 years and its estimated cost is £800 000, the average annual cost is £2000. If some measure can reduce the probability of the occurrence to once in 1000 years, this means a reduced average annual cost of £800. Should the cost of this measure, converted to an annual charge, be £1500 it would not be worth while since it would save only £1200. If, however, its annual charge were only £1000 the expenditure would be justified.

The same procedure could be adopted for situations involving risks to people if there were any agreed value placed on human life. There is, however, no such agreement and the usual procedure is to design for an agreed maximum level of risk. This is expressed as a fatal accident frequency rate (FAFR) or fatal accident rate (FAR). It is the number of fatal accidents in a group of 1000 people during their working life, usually taken as 10^8 man hours. The figure for the UK chemical industry is 4 if the Flixborough accident is ignored, and about 5 if it is included in a 10-year average. The figure for all premises covered by the UK Factories Act is also 4 (Green 1982). Since about half of the accidents in the chemical industry are unconnected with the material being handled, and involve falling down stairs, or vehicles, the FAFR for a chemical plant should be no more than 2.

This figure represents the total risk; where it is difficult to predict each individual risk it is suggested that the figure for any particular risk should not exceed 0.4. To give an idea of the import of this value, it is estimated that we all accept an FAFR of about 0.1 when engaged in driving, flying or smoking (Green 1982). This figure is somewhat different from that quoted elsewhere; one text (McCormick 1981) gives the FAFR for hunting, skiing and smoking as 10-100.

A detailed analysis depends upon a knowledge of the likely failure rates of the various components of the plant which are best obtained from previous experience with similar equipment. Typical figures quoted are as follows:

- The failure rate for natural gas pipelines in the USA is about 47×10^{-5} per mile per year.
- The rate for sudden failure of a pump (including cable, motor and gearbox) is about 0.4 per year.
- The rate for failure of a level controller is about 0.5 per year.
- The rate for a control valve failing shut is about 0.5 per year.

12.4 The oil industry

The oil industry is particularly susceptible to fire and explosion hazards since its raw material and nearly all of its products are flammable. Reliable plant designs for land-based equipment have been developed over many years, but the exploitation of the North Sea oilfields revealed several problems arising from the hostile environment which had not previously been encountered. Some of the oilfields were several hundred miles offshore where the sea is over 600 feet (180 m) deep and waves often 30-50 feet (9-15 m) high. The height of a wave expected once in a hundred years is over 100 ft (30 m) (Carson 1982). The lack of previous experience in working under these conditions meant that some structures failed; for example, in March 1980 the Alexander Keilland, an oil rig in the Norwegian sector, turned turtle in some 20 minutes with the loss of 123 lives. The cause was ascribed to the collapse of one of the five supporting columns. The column was held to the platform by six bracings and an opening had been cut into one of them to house a hydrophone positioning control which was welded in place. The reduction in strength caused fatigue fractures in the welds and the bracing, which eventually failed, so throwing extra load on to the other bracings which caused them to fail in turn. The column became detached from the platform, giving the rig a list of some 30°. The damage caused to the deck and the lack of compliance with instructions for watertight bulkheads and ventilators allowed much of the deck to be flooded and the whole structure turned over in some 20 minutes.

In this case the accident had a number of contributory causes: the effect of cutting an opening in a bracing had not been investigated fully, the spread of fatigue cracks in various welds and the structure itself had not been observed, and instructions about watertight doors and ventilators were ignored. In view of the harsh environment it is clear that some degree of redundancy should have been built into the structure so that it would survive if at least one, and preferably two, of the bracings failed. Also some interlocking mechanism could be provided which would prevent use of the rig in certain circumstances unless the watertight doors were closed.

By the mid-1970s the likelihood of a worker on an offshore installation in the British sector of the North Sea being killed was about 11 times greater than that of a construction worker and nearly six times greater than that of a miner. These figures do not include the 167 killed in the 1988 Piper Alpha explosion; 63 were killed in the period 1969-79, so the inclusion of the 167 would increase the 10-year average from 63 to 230, a factor of nearly four. This should preferably be included in the figures for 1979-89, but these are not yet available.

For over a century it has been realized that many aspects of reliability which affect the safety of industrial workers or the general public cannot be left to industry without government regulation and monitoring. Generally legislation is introduced some years after new practices or processes have been developed, often due to public alarm after some fatal accident. This is clearly evident in the history of the railways and coal mines, and is equally

144 *Reliability in Instrumentation and Control*

true of the offshore oil industry. Here matters are complicated because the platforms are often outside territorial waters and legislation was needed to extend government control (Continental Shelf Act 1964). This was passed in a great hurry and made little provision for safety measures. The Health and Safety at Work Act of 1974 made great improvements in the safety legislation in factories and other workplaces but was not extended to the Continental Shelf until 1977. A further complication not envisaged originally was that some investigations such as inquests would take place under Scottish law which has many differences from English law. An example of the legal difficulties arose in 1976 when the Grampian Police received a report that some fires which could have been started deliberately had occurred on a Panamanian-registered barge which was owned by a Dutch company and was on charter to an American company and operating within the safety zone of a production platform in the North Sea. At the time of the incident the barge was lying outside the safety zone because of bad weather. Although the police visited the barge, it was subsequently concluded that they had no jurisdiction (Carson 1982).

12.5 Reliability of oil supply

A typical oil platform contains a large number of separate units, not all of which contribute directly to the oil supply. In order to enhance the system reliability and to allow for routine maintenance much of the equipment is replicated. The overall reliability can then be estimated in terms of the reliabilities of the individual units, with allowance for any duplicate units.

Figure 12.1 shows a fault tree representing part of the pumping system of a typical oil platform having two supply paths. In the figure the symbols for unit reliability such as R_2, R_3 and R_6 are shown within the blocks which represent the units, and the reliabilities of the supply path to a particular point such as R_4 and R_5 are shown outside the blocks.

For path 1 to be operative, we require a supply of oil to the transfer pump and the transfer pump and the level control valve to be operating correctly. Thus the reliability of Path 1 is

$$R_4 = R_1.R_2.R_3$$

Production can be sustained if either Path 1 or Path 2 is working. Thus the reliability up to the final cooler is

$$R_5 = 2R_4 + (R_4)^2$$

assuming that both paths gave the same reliability.

Beyond this we need both the temperature control valve and the final oil cooler to be operative, so that the reliability of supply up to the storage facility is

Process and plant control 145

$R_9 = R_7.R_8 = R_5.R_6.R_8 = [2R_4 - (R_4)^2] R_5.R_6.R_8$

where

$R_4 = R_1.R_2.R_3$

On a large platform there may be four paths obtained by duplicating the equipment shown in Figure 12.1 which will give greater reliability if only one path need be operative.

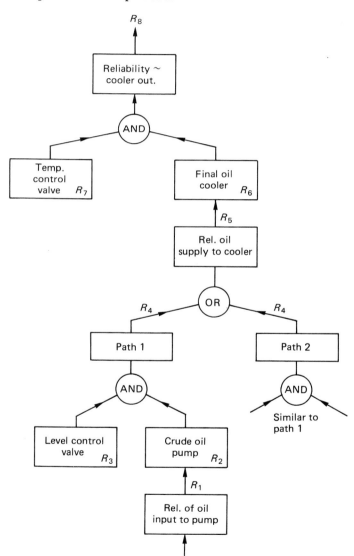

Figure 12.1 *Fault tree of part of oil platform*

12.6 Electrostatic hazards

In many chemical plants and in the transport and refinery of oil the hazard of igniting flammable vapours and gases is countered by flameproof enclosures and the segregation of potential spark-generating equipment from the flammable materials. Where the sparks or arcs are generated by current-carrying conductors this is a comparatively straightforward procedure, but in certain circumstances electrostatic potentials may be generated, and these are sometimes difficult to predict. The most likely cause is friction between insulators; under dry conditions potential of 10 kV or more can easily be generated. In conjunction with a capacitance of 50 pF (typical for a metal bucket) the stored energy is given by $E = 1/2\ C.V^2 = 2.5$ mJ.

This energy is sufficient to ignite hydrocarbons, solvent vapours and ethylene (Campden 1991). Figures quoted for minimum spark ignition energy (Handley 1977) are

Vapour – oxygen mixtures	0.002-0.1 mJ
Vapour – air mixtures	0.1-1.0 mJ
Chemical dust clouds	5-5000 mJ

Many plastic materials such as nylon are good insulators and readily generate static, so where sparks could be dangerous it is essential to earth all conducting objects in the vicinity.

Insulating liquids flowing in pipelines can also carry charge, generating currents of up to 10^{-6} A, and powders emerging from grinding machines can generate currents of 10^{-4} to 10^{-8} A. If we have a current of 10^{-7} A flowing into an insulated metallic container with a capacitance of 100 pF (e.g. a bucket), its potential will rise at a rate of 1 kV per second. In practice there will be some leakage and the rise in potential will be somewhat slower; it is nevertheless clear that potentials sufficient to cause ignition of flammable mixtures can be produced quickly.

Apart from earthing all metal objects in the working area, a number of other steps can be taken to reduce the risk of ignition.

1. Where flammable substances are being transported all vehicles and pipes used for the transfer must be bonded to earth before transfer starts.
2. Liquids having a very high resistivity can retain charge for some time even when passed through an earthed pipe. Their resistivity can be markedly reduced by introducing a few parts in a million of an ionic agent, so much reducing the hazard.
3. As the charging current generated by liquid flowing through a pipe is roughly proportional to the square of its velocity, the hazard can be reduced by ensuring that flow velocities are low.

4 Flammable atmospheres can be avoided by using an inert gas to dilute the concentration of flammable gas or vapour.
5 As static can be generated by free-falling liquids, entry and discharge pipes should be taken to the bottom of storage tanks, to avoid this.
6 In humid conditions insulators attract a conducting layer on their surface which provides paths for charges to leak away to earth. Thus static hazards can be much reduced by operating plant in air with a relative humidity greater than about 60%.
7 Most float-operated level indicators fitted to storage tanks use metal parts which must be firmly earthed.
8 People walking on synthetic carpets or flooring can easily become charged to a potential 10 kV or more. Before handling flammable liquids they should be earthed, for example by using conducting footwear and a conducting floor (BS 3187, BS 3389).

It is generally accepted that static electricity is likely to be generated only in a dry environment and high humidity is a valuable preventative. Surprisingly three explosions occurred in oil supertankers in 1969 while their tanks were being washed out with a high-pressure jet of sea water. After small-scale tests it was concluded that charges were liberated by friction between the water and the tank walls, and accumulated on falling water masses called 'water slugs', causing spark discharges between them (Campden 1991).

One hazard which has arisen only recently is the ignition of explosive and flammable atmospheres by radio transmissions. These occur in the vicinity of very high-power transmitters when the voltages induced in conductors are sufficient to create a spark or arc. Since the fields typical of normal broadcasting are only of the order of millivolts per metre it may be thought that cases of ignition should be very rare; however, with some high-power transmitters now delivering powers of a megawatt or more cases have occurred and it is recognized that oil refineries should not be located too close to high-power transmitters. BS 6656 and BS 6657 deal with this situation and suggest safe distances beyond which ignition should not occur.

12.7 The use of redundancy

As with electronic systems redundancy is a widely used method of improving the reliability of many industrial plants and processes. Since many of the units in these plants such as pumps, valves and compressors involve moving parts they are prone to wear and require periodic inspection and maintenance. Where the plant is required to operate continuously, isolating valves are required on either side of the unit so that it can be taken

148   *Reliability in Instrumentation and Control*

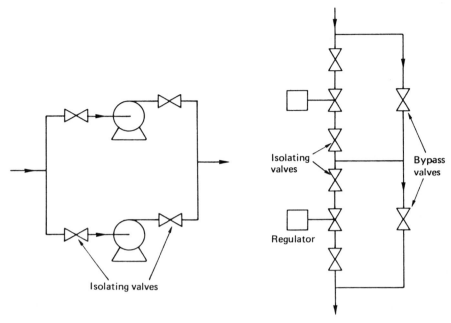

Figure 12.2 *Parallel and series redundancy*

off stream for inspection, maintenance or exchange. The arrangement depends upon the way in which redundancy is used.

Where a gas or liquid is pumped into a container and a failure of supply would be dangerous, two or more pumps may be installed in parallel, as shown in Figure 12.2, with isolating valves which allow each pump to be disconnected. Sometimes the reverse situation obtains when in an emergency the flow must be cut off. To allow for failure two or more valves can be installed in series; if they are motor operated, separate or standby power supplies may be desirable. In order to take a valve off stream it is necessary to provide a bypass as well as isolating valves, as shown in Figure 12.2.

Other devices may also be connected in series for greater reliability; for example, in some satellites pressure regulators are used for the propellant supply. Their main failure mode is to open, that is no regulation, so two regulators are used in series for better reliability.

Bibliography

Atallar, S. (1980) Assessing and managing industrial risk. *Chemical Engineering*, 8 September, p 94

Austin, W. M. and Dukek, W. G. (1983) *Electrostatic Hazards in the Petroleum Industry*, Research Studies, Letchworth

Bennett, P. A. (1988) The specification, testing and application of flameproof and increased safety electrical enclosures. *IEE Conference Proceedings*, No. 296
Bodurth, F. T. (1980) *Industrial Explosion Prevention and Protection*, McGraw-Hill, London
Butcher, E. G. and Parnell, A. C. (1983) *Designing for Fire Safety*, Wiley, Chichester
Campden, K. (1991) Electrostatics. *IEE Review*, October, p. 335
Carson, W. G. (1982) *The Other Price of Britain's Oil*, Martin Robertson, Oxford
Cohen, E. M. (1985) Fault-tolerant processes. *Chemical Engineering*, 16 September, p. 73
Danen, G. W. A. (1980) Electrical safety of process instrumentation in areas with potentially explosive gas atmospheres. *IEE Conference Proceedings*, No. 296
Dhillon, B. S. and Rayapati, S. N. (1988) Chemical-system reliability: A review. *IEEE Transactions on Reliability*, **R-37**, 199
Fussell, J. B. (1976) How to hand-calculate system reliability and safety characteristics. *IEEE Transactions on Reliability*, **R-24**, 169
Green, A. E. (ed.) (1982) *High Risk Safety Technology*, Wiley, Chichester
Gregory, R. J. and Brown D. R. (1988) The application of an industry code for the classification of hazardous areas. *IEE Conference Proceedings*, No. 296
Handley, W. (1977) *Industrial Safety Handbook*, 2nd edn, McGraw-Hill, Maidenhead
Henley, E. J. and Kumamuto, H. (1981) *Reliability Engineering and Risk Assessment*, Prentice Hall, Englewood Cliffs, NJ
Institute of Chemical Engineers (1990) Safety and loss prevention in the chemical and oil industries. *Symposium 1990*, Institute of Chemical Engineers, London
Jones, P. (1988) The development of the new Institute of Petroleum Code on area classification. *IEE Conference Proceedings*, No. 296
Kharbanda, O. P. and Stallworthy, E. A. (1988) *Safety in the Chemical Industry*, Heinemann Professional, London
King, R. W. (1990) *Safety in the Process Industries*, Butterworths, London
Lees, F. P. and Ang, M. L. (1989) *Safety cases within the control of Industrial Major Accident Hazards (CIMAH) Regulations 1984*, Butterworths, London
Marshall, V. C. (1987) *Major Chemical Hazards*, Ellis Horwood, Chichester
McCormick, N. J. (1981) *Reliability and Risk Analysis, Methods and Nuclear Power Applications*, Academic Press, London
Parker, R. J. (Chrmn) (1975) *The Flixborough Disaster. Report of the Court of Inquiry*, HMSO, London
Sansom, D. F. and Franklin, D. P. (1988) Certification to harmonized standards, *IEE Conference Proceedings*, No. 296
Sefton, B. (1989) Safety-related instrument systems for the process industry. *IEE Conference Proceedings*, No. 314
Strawson, H. (1973) Electrostatic fires and explosions. *The Chartered Mechanical Engineer*, **20**, 91
Towle, L. C. (1975) Instrumentation requirements in Zone 2 locations. *IEE Conference Proceedings*, No. 134
Townsend, C. A. (1988) Hazardous area classification – the final solution. *IEE Conference Proceedings*, No. 296
Wells, G. L. (1980) *Safety in Process Plant Design*, Godwin, London

British Standards

BS 787 Specification for mining type flame-proof gate end boxes. Parts 1-4 1968-72.
BS 889: 1965 (1982) Specification for flameproof electric lighting fittings.

BS 2915: 1960 Specification for bursting disc and bursting disc devices for protection of pressure systems from excess pressure or vacuum.

BS 3187: 1978 Specification for electrically conducting rubber flooring.

BS 3395: 1989 Specification for electrically bonded rubber hoses and hose assemblies for dispersing petroleum fuels.

BS 4137: 1967 Guide to the selection of electric equipment for use in division 2 areas.

BS 4200 Guide on reliability of electronic equipment and parts used therein. Parts 1-8, 1967-87.

BS 4683 Specification for electrical apparatus for explosive atmospheres. To be replaced by BS 5501.

BS 4778 Quality vocabulary. Part 1: 1987 International terms. Part 2: 1979 National terms.

BS 4891: 1972 A guide to quality assurance.

BS 5000 Rotating electrical machines of particular types or for particular applications. Parts 1-17.

BS 5345 Code of practice for the selection, installation and maintenance of electrical apparatus for use in potentially explosive atmospheres (other than mining or explosive manufacture and processing). Parts 1-8 1978-90. Also see EN 50014-20 (IEC 79), EN 50028 and EN 50039.

BS 5420: 1977 (1988). Specification for degrees of protection of enclosures of switchgear and control gear for voltages up to 1,000 V a.c. and 1,200 V d.c. Now superseded by BS EN 60947-1: 1992.

BS 5501 Electrical apparatus for potentially explosive atmospheres. See also EN 50014 and EN 50020.

BS 5750 Quality systems. Parts 0-6 1981-7

BS 5760 Reliability of constructed or manufactured products, systems, equipments and components. Parts 0-4 1981-6.

BS 6132: 1983 Code of practice for safe operation of alkaline secondary cells and batteries.

BS 6133: 1985 Code of practice for the safe operation of lead and secondary cells and batteries.

BS 6387: 1983 Specification for performance requirements for cable required to maintain circuit integrity under fire conditions.

BS 6467 Electrical apparatus with protection by enclosure for use in the presence of combustible dusts. Parts 1 and 2. 1985 and 1988.

BS 6656: 1986 Guide to the prevention of inadvertent ignition of flammable atmospheres by radio-frequency radiations.

BS 6657: 1986 Guide for prevention of inadvertent initiation of electro-explosive devices by radio frequency radiation.

BS 6713 Explosion prevention systems. Parts 1-4: 1986.

BS 6941: 1988 Specification for electrical apparatus for explosive atmospheres with type of protection 'N'. Replaces BS 4683: Part 3.

BS 9400: 1970 (1985) Specification for integrated electronic circuits and micro-assemblies of assessed quality.
BS 9401-94 deal with detail specifications for particular forms of integrated circuit.

British Standard Codes of Practice

BS CP 1003 Electrical apparatus and associated equipment for use in explosive atmospheres of gas or vapour other than mining applications. Largely replaced by BS 5345.
BS CP 1013: 1965 Earthing.
BS CP 1016 Code of practice for use of semi-conductor devices. Part 1: 1968 (1980) General considerations. Part 2: 1973 (1980) Particular considerations.

European and harmonized standards

BS QC 16000-763000 Harmonized system of quality assurance for specific components.
BS CECC 00009-96400 Quality assessment of specific classes of component.
BS E9007: 1975 Specification for harmonized system of quality assessment for electronic components. Basic specification: sampling plans and procedures for inspection by attributes.
BS E9063-377 deal with specific classes of component.

British Standards are available from the BSI Sales Dept, Linford Wood, Milton Keynes MK14 6LE.

Index

Accelerated life tests, 31
Ada, 90, 99
ADC, 4, 6, 42, 106
AGR plant, 127, 130
Aircraft control systems, 116–19
Algol, 97
Amplifier, redundant, 63
Analogue:
  coding, 41–3
  current scaling, 4
  I.C.s, 30
  signals, 3, 4, 115
  voltage scaling, 4
Automation, 67
Arrhenius equation, 27
Availability, 15, 16

Bath-tub curve, 23
Blind landing system, 83
Bridge, Wheatstone, 6
British Standard codes of practice, 151
British Standard Institute, 56, 57
British Standard specifications, 149–51
Built-in testing, 50, 122
Burn-in, 24

CAMAC, 5
Capacitors, 48
CASE, 91
Chemical plant hazards, 139–42
Chernobyl power station, 77, 80, 125
Circuit fault analysis, 84–6
CMOS circuits, 30, 108
Coding, analogue, 41–3
Common mode faults, 64–6, 118, 119, 132–4
Common mode voltage, 106
Component screening, 31, 32
Component selection, 56, 57
Component temperature, 28–30
Compound systems, 18–21
Confidence level, 32
Confidence limits, 32

Control:
  digital, 3
  room, 2
  system faults, 82–4
  system nuclear reactor, 125–36
  systems, 1, 2
Coral, 90
CRT displays, 7

DAC, 4, 43, 106
Data highways, 5
Data transmission, 105–11
  analogue, 106, 107
  digital, 108
Data validation, 107
Derating of components, 28
Design:
  automation, 49
  errors, 49, 50
  of controls, 71
  system, 41, 44
  user friendly, 71
  worst-case, 45
Digital coding, 43, 115
Displays, 7
  visual, 75
Diversity reception, 64
Down-time, 16
Duplicate redundancy, 113, 117

Electrostatic hazards, 146, 147
Environmental testing, 38
Error:
  correction, 108, 110
  detection, 108, 109
  fail-dangerous, 79, 80, 129, 130
  fail-safe, 79, 80, 129
  in data transmission, 105
ESONE, 5
European Standards, 151

FAFR, 142
Fail-dangerous errors *see* Error, fail-dangerous

154 Index

Fail-safe errors *see* Error, fail-safe
Fail-safe systems, 80, 82, 131
Failure:
  catastrophic, 25, 26
  degradation, 25, 26
  density, 89, 90
  density function, 24, 34
  intensities, 89, 90, 92
  law, exponential, 13–15
  modes, 25
  open-circuit, 84–6
  short-circuit, 84–6
Failure rate, 12, 24
  chemical plant, 142
  component, 23
  effect of temperature, 26–8
  effect of voltage, 30
  system, 35
Failure-tolerant systems, 100, 101
FAR, 142
Fault:
  analysis of circuits, 84–6
  common-mode, 64–6
  density, 89, 90
  estimation, 94, 95
Faults and failures, software, 88, 89
Faults, software, 95
FIT, 60, 85, 86
Flameproof enclosures, 139, 140
Flash point, 140
Fortran, 97, 98
Frequency diversity, 111

GPIB, 5
Gypsy, 100

Hamming codes, 109
Hazardous areas, 141
Health and Safety at Work Act 1974, 144
Heat sink, 29
High reliability software, 90, 91
Human errors, 68
Human operator, 68–70
  response time, 70

Inductors, 48
Infant mortality phase, 23
Instruments, 1, 2
  layout, 73–5
  pointer type, 8
Intrinsic safety, 140

Job specification language, 99, 100

Laddic, 135
Languages, job specification *see* Job specification language
LCD, 7, 8
LED, 8, 73, 75
Life tests, accelerated, 31
Longitudinal voltage, 106

Magnox nuclear reactor, 127, 131, 134
Majority voting, 59, 60, 116, 122, 127, 128, 135
Margin, performance, 44
Microprocessor, 2, 42, 50, 89, 131
MIL-HDBK-217B, 80
MTBF, 12, 23, 55, 56, 59, 116–19
MTTF, 12

Newspeak, 99
Nuclear reactor control system, 125–36

Optical fibre data transmission, 106

Parity bit, 108, 109
PIUS, 126
Plasma panel, 7
Pneumatic transmission, 4
Poisson probability distribution, 13
Power-assisted steering, 83, 84
Printer, 10
PRISM, 126
Programming, structured, 96, 97
Program testing, 101
Protection system:
  relay, 81, 131–3
  static, 134–6

Race hazards, 51, 52
Railway braking systems, 120
Railway control, 119–21
Railway signalling, 119–21
Recording:
  chart, 9
  magnetic, 9
  techniques, 9
Redundancy, 57–64
  analogue, 62
  duplicate, 113, 117
  in power supplies, 65
  in data transmission, 110, 111
  in process plant, 147–8
  level, 61
  triplicate, 116–18
Relay properties, 81

Index

Relay safety systems, 131–3
Relay tripping, 81
Reliability:
  and MTBF, 12
  budgets, 81
  compound systems, 18–21, 36
  definition, 11, 12
  duplicate, 21
  growth, 91–4
  growth models, 91–4
  oil supply, 144, 145
  optimum, 17
  parallel systems, 36
  radio transmitters, 113–15
  software, 87–102
  triplicate systems, 21, 36
Resistors, metal film, 47
Robotic systems, 122

Safety:
  in oil industry, 143, 144
  margins, 44
  procedures, 76, 77
Satellite links, 115
Screening of assemblies, 33
Screening of components, 31, 32
Seeding of software faults, 94
Semiconductors, effect of, 48
Signal coding, 41
Sneak circuits, 51
Space diversity, 111
SSADM, 97
Steering, power-assisted *see* Power-assisted steering
Structured programming *see* Programming, structured

System:
  compound *see* Compound systems
  design *see* Design, system
  fail-safe *see* Fail-safe systems
  failure-tolerant *see* Failure-tolerant systems

Temperature, effect on components, 47–9
Testing, environmental, 38
Thermal resistance, 29
Three Mile Island power station, 76, 125
Tolerances:
  component, 46, 47
  parameter, 44, 45
Transducers, analogue, 106
Transmission, data, 2
Transverse voltage, 106
TTL circuits, 43

UPS, 64, 65
Up-time, 16

VIPER, 50
Voltage amplifier, 47
Voltage transients, 114
Voting circuit, 62
Voting, majority *see* Majority voting

Wear-out phase, 23, 24, 33–5
Worst-case design *see* Design, worst-case

Z (programming language), 100

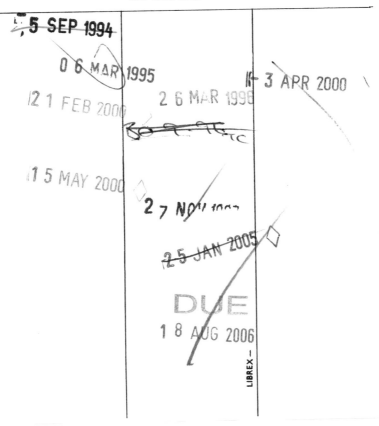